Foundations of Organic Chemistry: Worked Examples

Michael Hornby and Josephine Peach

Series sponsor: AstraZeneca

AstraZeneca is one of the world's leading pharmaceutical companies with a strong research base. Its skill and innovative ideas in organic chemistry and bioscience create products designed to fight disease in seven key therapeutic areas: cancer, cardiovascular, central nervous system, gastrointestinal, infection, pain control, and respiratory.

AstraZeneca was formed through the merger of Astra AB of Sweden and Zeneca Group PLC of the UK. The company is headquartered in the UK with over 50,000 employees worldwide. R&D centres of excellence are in Sweden, the UK, and USA with R&D headquarters in Södertälje, Sweden.

AstraZeneca is committed to the support of education in chemistry and chemical engineering.

OXFORD
UNIVERSITY PRESS

OXFORD

UNIVERSITY PRESS

Great Clarendon Street, Oxford OX2 6DP

Oxford University Press is a department of the University of Oxford.
It furthers the University's objective of excellence in research, scholarship,
and education by publishing worldwide in

Oxford   New York

Athens   Auckland   Bangkok   Bogotá   Buenos Aires   Calcutta
Cape Town   Chennai   Dar es Salaam   Delhi   Florence   Hong Kong   Istanbul
Karachi   Kuala Lumpur   Madrid   Melbourne   Mexico City   Mumbai
Nairobi   Paris   São Paulo   Shanghai   Singapore   Taipei   Tokyo   Toronto   Warsaw
with associated companies in   Berlin   Ibadan

Oxford is a registered trade mark of Oxford University Press
in the UK and in certain other countries

Published in the United States
by Oxford University Press Inc., New York

© G. M. Hornby and J. M. Peach
The moral rights of the author have been asserted
Database right Oxford University Press (maker)

First published 2000

A catalogue record for this book is available from the British Library

Library of Congress Cataloging in Publication Data
(Data applied for)
ISBN 0 19 850583 3
Typeset by EXPO Holdings, Malaysia
Printed in Great Britain
on acid-free paper by The Bath Press, Bath

Series Editor's Foreword

Foundations of organic chemistry: worked examples is the companion Primer to the internationally highly acclaimed *Foundations of organic chemistry* (OCP 9). Together these Primers are designed to facilitate the transition from Chemistry studies at school to those at University by providing a concise and elementary coverage of core topics both descriptively and by way of worked examples. These Primers have been written in a very user friendly format by two Master chemistry teachers and will stimulate young people reading chemistry at advanced level in school, and serve to excite still further the interest of those students just starting their apprenticeships in chemistry at University.

Professor Stephen G. Davies
The Dyson Perrins Laboratory
University of Oxford

Preface

This book has been written in response to requests from readers of our *Foundations of organic chemistry*.* Practice in solving problems is the one certain way of ensuring a good working understanding of the basic principles of organic chemistry. Much of the pleasure in studying the subject lies in working out the often beautiful architecture of the molecules, in understanding their reactions and in devising elegant ways of making them. We have tried to share our enjoyment of problem-solving whilst helping students to reinforce and extend what they have learned at school.

We have retained the pattern of *Foundations*, starting with three chapters of questions and answers on basic physical organic chemistry and general mechanisms before going on to the mechanisms in more detail. We have not included spectroscopic problems (these are available in the companion volume, *Foundations of physical chemistry: worked examples*) but recognition of functional groups is emphasised here, as this is probably the most important aspect of such problems at this stage. We have tried again to keep the majority of ideas and reactions within reach of Advanced Level chemistry students, whilst also using some less familiar and more interesting compounds in our examples.

We are most grateful to all those who have given us valuable criticism and advice: in particular Peter Carpenter (Roedean School), Martin Grossel (Southampton University), Jean Johnson (Open University) and Stephen Lunt (Bradfield College). We remain indebted to those others who helped us with *Foundations*: John Nixon, David Smith and Sydney Bailey. Professor Stephen Davies has given us unfailing support in the production of both books. Finally we would like to thank Yuki Soga for her help with the diagrams.

We would like to dedicate this book to Janet and to John, to Sarah and William, and to Charles, Sarah and James.

Buckingham   G. M. H.
Oxford   J. M. P.

*Oxford Chemistry Primer No. 9, Oxford University Press, 1993 referred to throughout as H & P

How to use this book

1. READ, LEARN AND REVISE THE CHEMISTRY SUGGESTED *before* you attempt to answer the questions in any particular chapter. It often helps to use more than one book, if possible.
2. COVER THE ANSWER (below the dotted line) with a sheet of paper *before* you start reading the question.
3. READ THE WHOLE QUESTION SLOWLY. Don't just pick out key words and hope that you know what the question is going to be. Hints are sometimes given in the margin beside the question.
4. WRITE THE WHOLE OF YOUR ANSWER AND CHECK IT *before* you look at the book's answer.
    Check your drawings. Do they represent the correct compounds? Are the normal rules of valency observed?
    Check your equations. Numbers of atoms and charges must balance, and numbers of electrons too.
    Check your calculations, especially any molecular formulae and relative molecular masses used. Check that your answer is roughly what you expect: for example that the pH of an aqueous base is greater than 7.
    Check over your answer—is it chemically reasonable and is it compatible with all the data given in the question?
5. COMPARE YOUR ANSWER WITH THE BOOK'S. If the two answers say essentially the same thing, but not necessarily in exactly the same words, that is good. If they differ markedly, try to work out why. Comments in the margin beside the answers include alternative explanations or further extensions, as well as some common errors.
6. HELP IS AT HAND! References to the relevant sections in Hornby and Peach, 'Foundations of Organic Chemistry' are given at the end of each question, and there are similar sections in all good textbooks. Your teacher or tutor will always be able to help you.

Contents

1 Molecules

Questions by topic

1.1 What you need to read about or revise for this chapter

A Atoms, molecules and bonding: electronic configurations of atoms and ions especially in the periods Li to Ne and Na to Ar. Dot diagrams of atoms, ions and molecules. Prediction of shapes using electron-pair repulsion theory.
Ionic and covalent bonding; dipoles. Hybridisation: single, double and triple bonding by carbon and other elements. Electron delocalisation (resonance).

A: 1, 2, 3, 4, 5, 6, 7, 8, 9

B Drawing and naming molecules: to include skeletal diagrams. Interconversion of names and diagrams; the common functional groups.

B: 10, 11, 12, 13

C Isomerism and stereochemistry. Structural and stereoisomerism: geometrical (*cis/trans*) and optical isomerism. Chirality, planes and centres of symmetry.

C: 14, 15, 16, 17, 18, 19, 20

D Intermolecular forces. van der Waals forces, dipolar interactions, hydrogen bonding: effects on physical properties.

D: 21, 22, 23, 24, 25.

1.2 Worked examples

Question 1

XH_4 species can carry positive, negative or zero charge, for example NH_4^+, BH_4^- and CH_4. Explain the charge carried by each of these.

Draw dot diagrams for the formation of BH_4^- and NH_4^+.

Answer

The outer shell electrons of the atoms are:
H $1s^1$ B $2s^2 2p^1$ C $2s^2 2p^2$ N $2s^2 2p^3$

Combining one carbon and four hydrogen atoms:

$$C + 4H \rightarrow CH_4 \quad \text{(All species are uncharged.)}$$

Similarly $B + 3H \rightarrow BH_3$ and $N + 3H \rightarrow NH_3$.

Formally, BH_4^- and NH_4^+ can be derived as follows:

$$BH_3 + H^- \longrightarrow BH_4^-$$

i.e.

Here is a different way of working out the charges. From the dot diagram (see left), BH_4^- has eight outer shell electrons. The four H atoms contribute one each, leaving four for the B: one more than a B atom, so B is negatively charged. NH_4^+ has eight too; four from four H atoms leaves four for the N; one less than an N atom, so the N is positively charged.

$$NH_3 + H^+ \longrightarrow NH_4^+$$

i.e.

The pair of electrons for the new bond in BH_4^- comes from H^-. Formally, the B atom in BH_4^- has acquired a half-share in the new bonding pair, that is, it has gained one electron and so becomes negatively charged. Conversely, ammonia's non-bonded pair (both electrons belonging to nitrogen) becomes a bonding pair, shared between N and H, so N becomes positively charged. Note: The *charges* in an equation must balance, as well as the atoms.

H & P section 1.3 (p 9)

Question 2

You may find dot diagrams helpful.

Explain the changes of shape which occur across the following reaction:
$BF_3 + NH_3 \longrightarrow F_3B \leftarrow NH_3$.

Answer

The product can also be drawn with charges

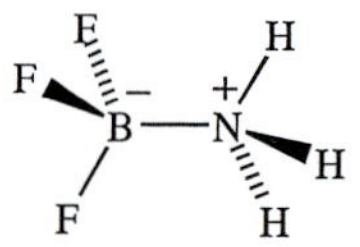

instead of with the ← dative bond. Compare this shape with the shape of ethane, H_3C–CH_3: they are very similar.

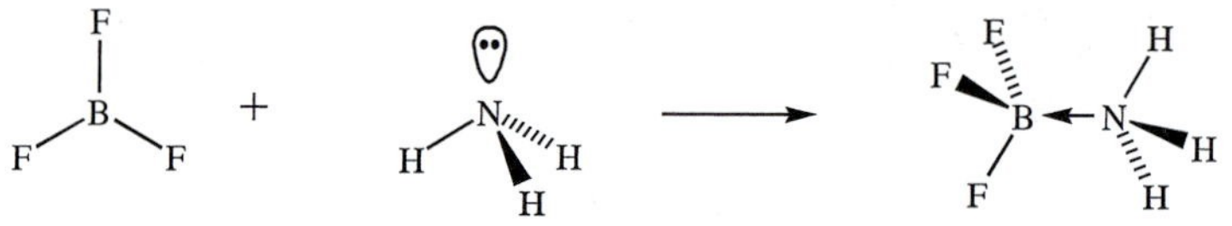

flat, trigonal

pyramidal (tetrahedral including e pair)

both B and N tetrahedral

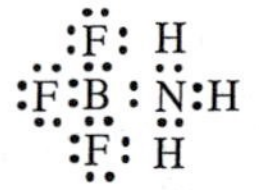

B has six outer shell electrons. These three pairs are at maximum separation in space, giving a trigonal shape with bond angles of ~120°. The B can accept another electron pair.

N has eight outer shell electrons. Four pairs at maximum separation gives a tetrahedral distribution.

Both B and N now have eight outer shell electrons, so both have a tetrahedral arrangement of bonds.

H & P section 1.3 (pp 9, 11)

Question 3

β-CAROTENE is a natural fat-soluble yellow dye. It is found in leafy green vegetables, grass and carrots. It is used commercially to colour margarine and smoked fish.

(a)   Draw out the **full** structure of β-carotene, **C**, to show all the carbon and hydrogen atoms.

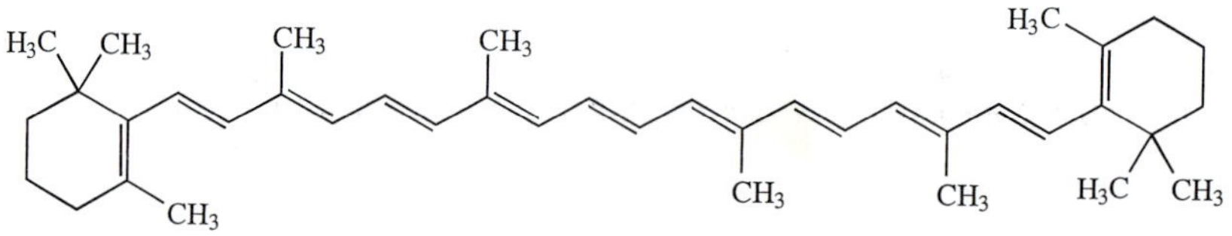

C

(b)   What is the empirical formula of β-carotene?

(c)   What method would you use to determine its molecular formula?

(d) β-Carotene is a natural product which is formed from the C_5 units of 2-methyl-1,3-butadiene, **D**.

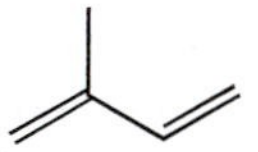

D

Using a diagram, show how eight C_5 units of **D** fit together to make the carbon framework of β-carotene.

Answer

(a)

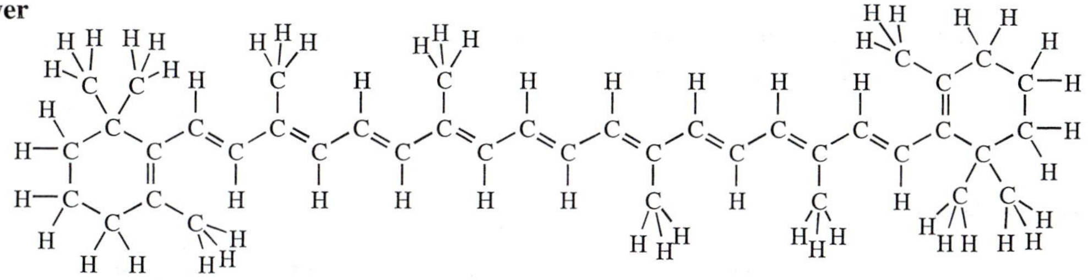

β-carotene

(b) C_5H_7, from the molecular formula, $C_{40}H_{56}$.

(c) Mass spectrometry.

(d)

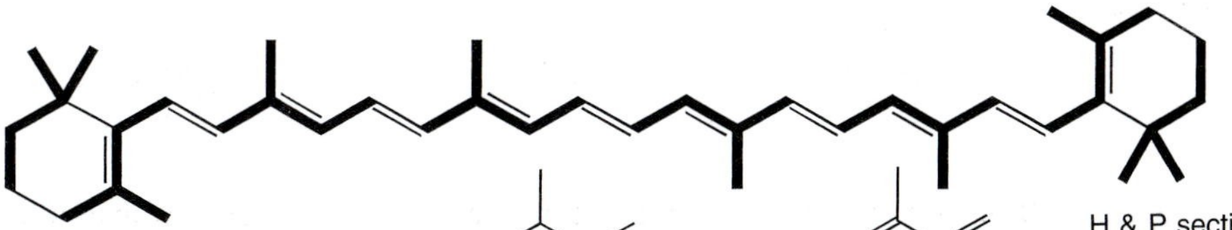

H & P section 1.3 (p 8)

Each section must have the five-carbon 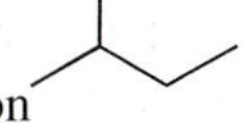 framework of 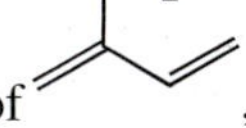 , **D**.

Question 4

(a) Draw dot diagrams of $(CH_3)_3CCl$ and $(CH_3)_3C^+$, using CH_3 for the methyl groups.

(b) Predict the shape of $(CH_3)_3C^+$ and its C–C–C bond angles, giving your reasoning.

Write down the equation for the ionisation of $(CH_3)_3CCl$.

Answer

(a)

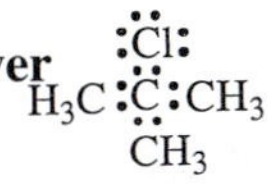

and

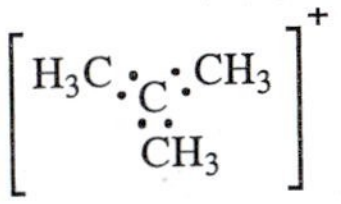

This shows how much more crowded $(CH_3)_3CCl$ is than $(CH_3)_3C^+$.

from

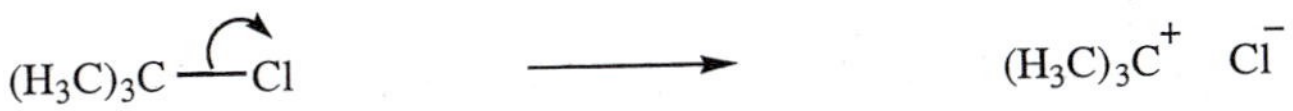

$\longrightarrow$ $(H_3C)_3C^+$ Cl^-

or

$\longrightarrow$

The charges in an equation must balance, and so must the total number of electrons. $(CH_3)_3C^+$ is electron-deficient because the central carbon has only six electrons.

(b)   Trigonal planar, angle ~120°.

The three pairs of electrons in the outer shell of the central C keep as far away from each other as possible to give the flat, symmetrical trigonal shape—similar to the planar BF_3 molecule.

H & P section 1.3 (p 9)

Question 5

(a)   Draw clear diagrams to show the shapes of the following molecules, explaining your answers.

Draw the bonding for each molecule and consider the shape at each atom (except H).

(i) CH_3OCH_3 (ii) CH_3COCH_3
(iii) $POCl_3$ (iv) H_2CCHCN

(b)   Predict shapes for these ions:

It may help to draw H_2SO_4 and then think how SO_4^{2-} is formed from H_2SO_4.

(i) SO_4^{2-} (ii)   H   Br

....

Answer

(a)   (i)

H_3C — O — CH_3

Tetrahedral (sp^3) carbon atoms. Angular around oxygen.

(ii)

H_3C — C — CH_3 (with O double bonded)

The central carbon atom is trigonal (sp^2).

(iii) $O=P$ with Cl, Cl, Cl

The four groups of bonding electrons are as far away from each other as possible.

Approximately tetrahedral around phosphorus, which has expanded its octet of outer electrons to 10.

(iv) $H_2C{=}CHC{\equiv}N$

The two C=C carbons are trigonal (sp^2) and the CCN is linear (sp).

(b) (i)

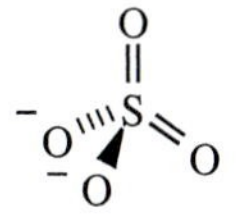

tetrahedral

The dot diagram for the bonding electrons is

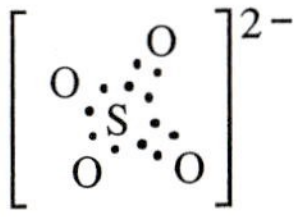

SO_4^{2-} has all its S–O bond lengths identical. It is delocalised and stabilised by resonance.

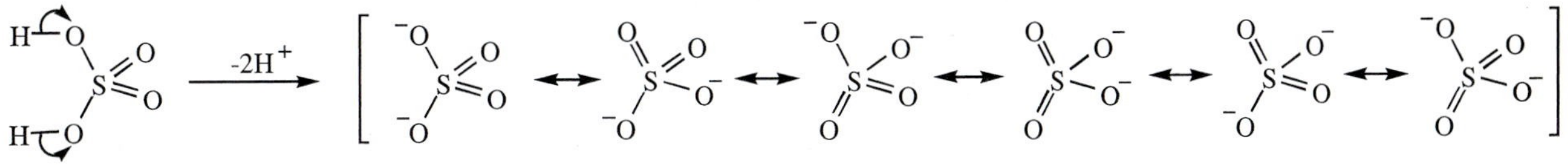

or

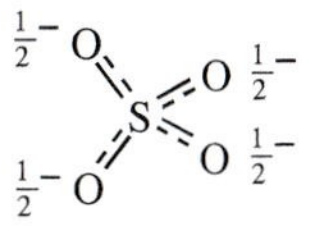

S has expanded its octet to 12 outer shell electrons in SO_3, H_2SO_4 and the ions SO_4^{2-} and HSO_4^-.

There are *no* nonbonded pairs on the S.

(ii)

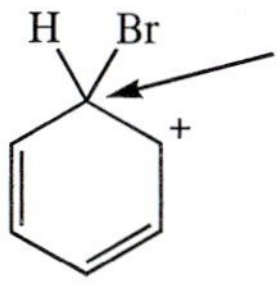

Tetrahedral C: all the others are trigonal planar

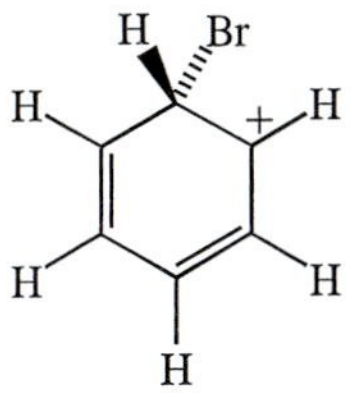

or

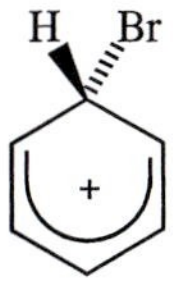

from

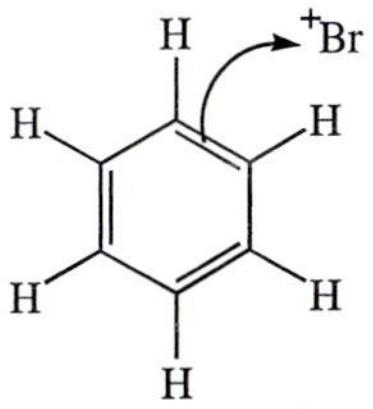

H & P sections 1.3, 1.5 (pp 7, 11)

Question 6

(a) Draw a diagram to show the resonance structures for the delocalised ethanoate ion, $CH_3CO_2^-$.

(b) Estimate its C–O bond lengths. The average C–O single bond length is about 0.14 nm and the average C=O double bond is about 0.12 nm.

....

Answer

(a)

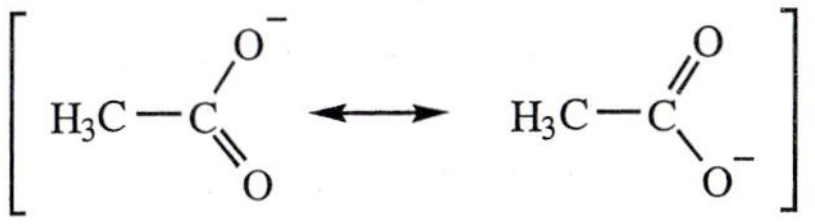

All resonance structures for a given compound must carry the same overall charge.

(b) About 0.13 nm; intermediate between C–O and C=O lengths.

Make the bond angles around the central C approximately trigonal, ~120° (*not* 90°).
Doubleheaded arrows ↔ are used ONLY for resonance structures. Note the difference from equilibrium arrows ⇌.

H & P section 1.3 (p 5)

Question 7

Draw a structural formula and a dot diagram for each of the following species: HNO_2, HNO_3, CH_3NO_2, NO_2^-, NO_3^- and NO_2^+.

....

Answer

HNO_2

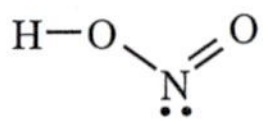

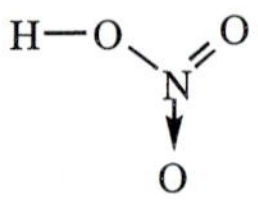

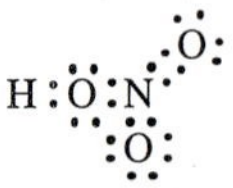

The $-NO_2$ parts of the rest of these are delocalised; these structural formulae and dot diagrams represent one possible arrangement of the electrons for each compound.

HNO_3

or

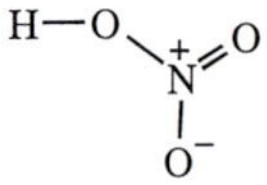

CH_3NO_2

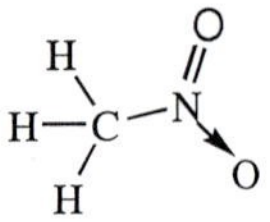

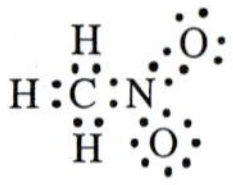

You can make the NO_2^- ion by taking a proton away from HNO_2:

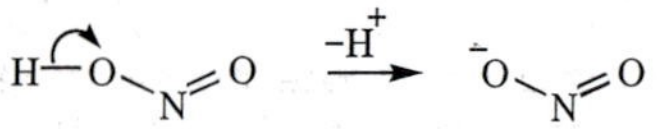

and similarly for the dot diagram. In the same way,

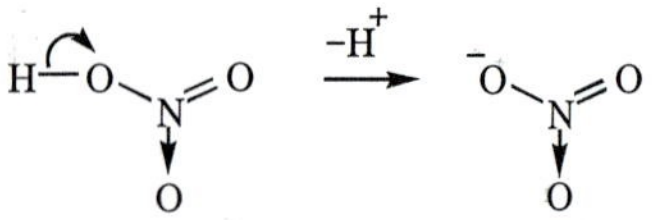

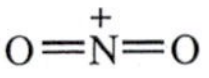

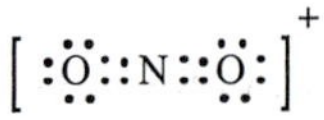

H & P section 1.3 (pp 7, 9)

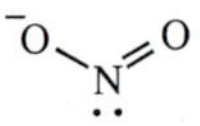

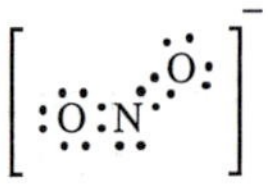

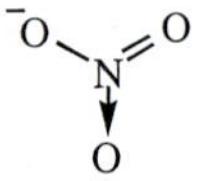

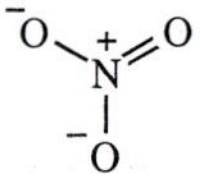

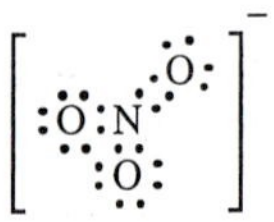

NO_2^-

NO_3^-

or

NO_2^+

$O{=}\overset{+}{N}{=}O$

Question 8

Draw delocalised (resonance) structures for NO_2^- and NO_3^-.

....

Answer

These delocalised structures can be related to each other by curly arrows.

NO_2^-

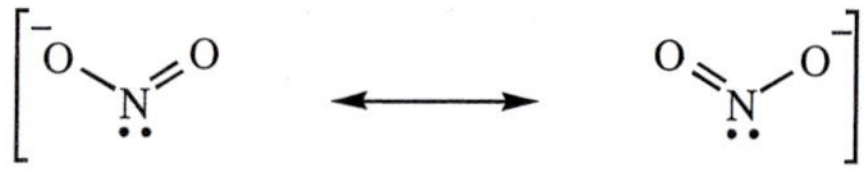

Two identical structures   Overall:

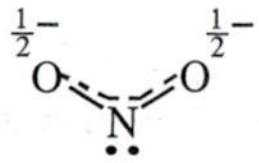

NO_3^-

$$\begin{bmatrix} \overset{-}{O}\diagdown\overset{+}{N}\diagup O & \longleftrightarrow & O\diagdown\overset{+}{N}\diagup \overset{-}{O} & \longleftrightarrow & \overset{-}{O}\diagdown\overset{+}{N}\diagup \overset{-}{O} \\ {\overset{|}{O}}^- & & {\overset{|}{O}}^- & & {\overset{\|}{O}} \end{bmatrix}$$

Three identical structures   Overall:

$$\overset{\frac{1}{3}-}{O}\diagdown\overset{+}{N}\diagup O^{\frac{1}{3}-}$$
$$O\,\tfrac{1}{3}-$$

H & P section 1.3 (pp 5, 7)

Question 9

How many outer shell electrons are there (shared plus nonbonded) on each of the N and O of coccinellin? Indicate how many nonbonded pairs there are on each of these two atoms.

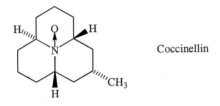

Coccinellin

Draw a dot diagram of the central N → O section.

COCCINELLIN is a 'defence compound' produced by ladybirds when attacked.

N-oxides can be drawn as

$>N \rightarrow O$ or as $>\overset{+}{N} - \overset{-}{O}$

$>N: \overset{..}{\underset{..}{O}}: \longrightarrow >N : \overset{..}{\underset{..}{O}}:$

....

Answer

N: eight electrons in outer shell, no nonbonded pairs.
O: eight electrons in outer shell, three nonbonded pairs.
The N has three single bonds to C and one dative covalent bond to oxygen, formed by sharing its original nonbonded pair with an oxygen atom, making four shared pairs, none nonbonded. The O has one shared pair, in the bond to N.

A dative bond, once formed, is an ordinary sigma bond. For example, all four NH bonds in NH_4^+ are identical.

H & P section 1.3 (p 2)

Question 10

Draw full structural formula showing every atom and bond, for the following molecules:

(i)   2-methylpropane;   (ii)   propanal;
(iii) ethyl methanoate;   (iv) ethanonitrile;
(v)   2-chlorobutane.

See end papers for functional groups.

....

Answer

(i)

$$C^3 - C^2 - C^1 \quad \text{with} \quad \overset{CH_3}{\underset{2}{|}}$$

2 - methyl propane

methyl on second C CH_3 group   3C skeleton

'2-methyl' means that there is ONE methyl group on C2.
It is a common error to think that 2-methyl means that there are two methyl groups on the C_3 chain. 'Dimethyl' is used for two methyl substituents, trimethyl for three and so on.

(ii)

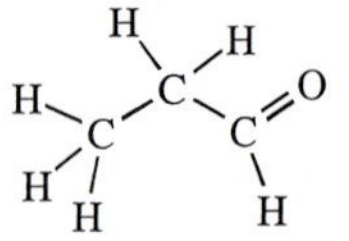

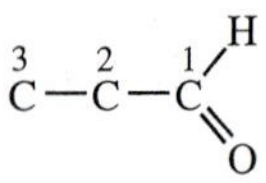

propan|al

3C skeleton   aldehyde group at end

An ester is conventionally divided:

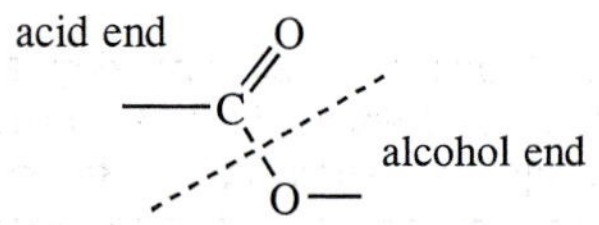

(iii)

Drawing methanoic acid as CH_3COOH (which is *ethanoic* acid, with a two-carbon chain) is a common mistake.

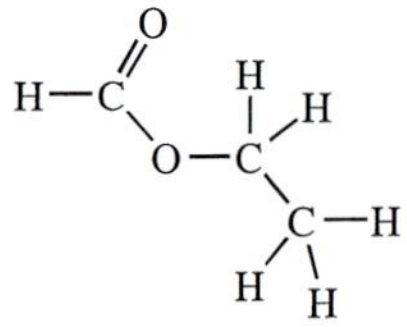

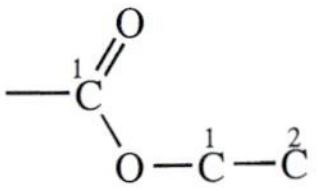

ethyl methan|oate

2 C at 'alcohol'   1C at 'acid'   ester
end of ester   end

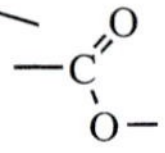

The C–C≡N bond angle is 180° because the central C is sp hybridised (like an alkyne).

(iv)

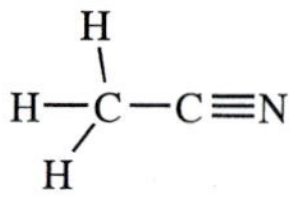

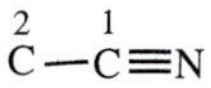

ethano|nitrile

2C skeleton   ≡N group at end

Number the C atoms to give the substituent on the lowest number. This is 2-chlorobutane rather than 3-chlorobutane.

(v)

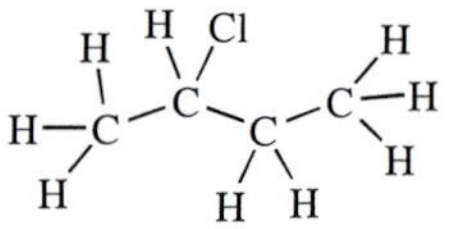

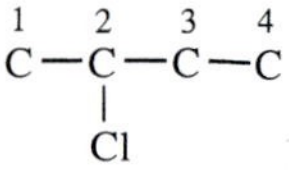

2-chloro| butane

one Cl on C -2   4C skeleton

H & P section 1.3 (p 7).

Question 11

Name the following compounds:

(i)

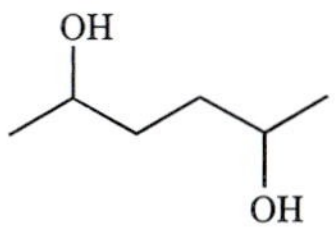

(ii)

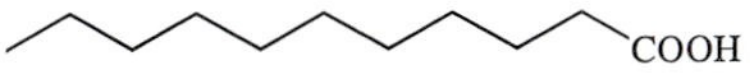

Always start by numbering the longest unbranched carbon chain, e.g. in (ii) there are 11 Cs so it is undeca-. Skeletal diagrams are explained in H & P (p8).

(iii)

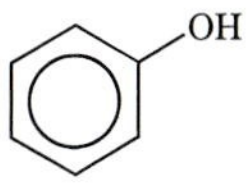

(iv)

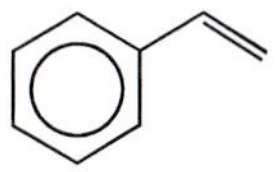

(v)

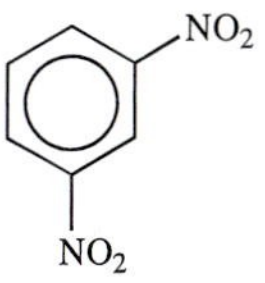

(vi)

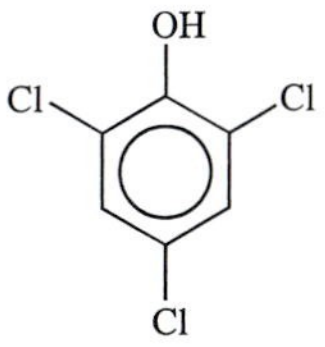

The C atoms of an aryl ring are numbered starting from a carbon carrying a substituent. If one substituent can be part of the name (e.g. phenol) then this is designated as position 1.

....

Answer

(i) 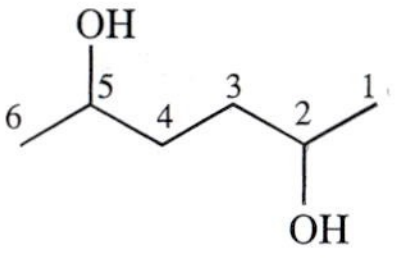 2,5-dihydroxyhexane or hexane-2,5-diol

(ii) 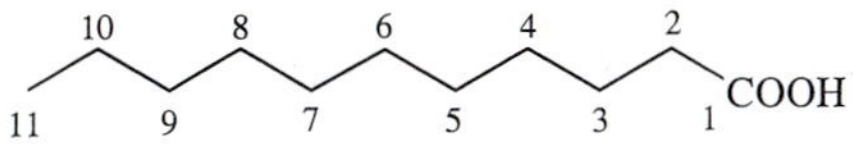 undecanoic acid

(iii) Phenol

(iv) Phenylethene

Phenylethene may be more familiar as styrene.

(v) 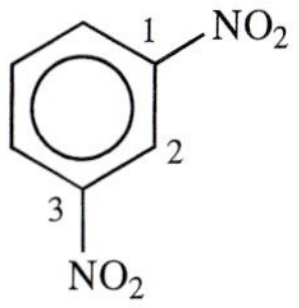 1,3 dinitrobenzene

Use the lowest numbers round the aryl ring, so (v) is 1,3- and not 2,4- or 3,5-.

The –OH has been incorporated into the name 'phenol', so counting starts from its carbon as 1.

H & P section 1.3 (p 7)

(vi)

2,4,6-trichlorophenol

Question 12

ASPARTAME is 200 times as sweet as sucrose and is much used in reduced-calorie drinks. It was discovered by chance, by a chemist working on a quite different problem.

Pick out the carboxylic acid and ester groups in aspartame, an artificial sweetener. Draw out the structure to show **all** the multiple bonds.

Aspartame

Carboxylic acids can be drawn

as

esters as

or any other way up! You must be able to recognise functional groups written any way round, especially when written in shorthand forms, e.g. CH_3COOCH_3, $CH_3CO_2CH_3$ and CH_3OCOCH_3 are all methyl ethanoate.

H & P section 1.3 (p 8) and endpapers.

Answer

carboxylic acid

ester

Don't forget to draw out the C_6H_5.

Question 13

Draw the molecules out to show all multiple bonds, and use the functional groups shown on the end papers to help you.

MENTHOL smells minty and has a cooling effect. It has been isolated from many types of mint plants and is also a constituent of rose oil.

Identify and name the functional groups in the following compounds.

Menthol

Aspirin

CHOLESTEROL is synthesised in the body and its metabolism is regulated by a specific set of enzymes. It is the main constituent of human gallstones, and is connected with the hardening of arteries.

Cholesterol

Muscarine cation

Answer

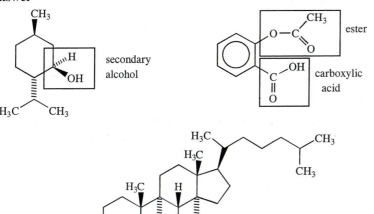

secondary alcohol

ester

carboxylic acid

alkene

secondary alcohol

secondary alcohol

tetra-alkyl ammonium ion

ether

More usually esters are written

not

The order in which the atoms are connected (their connectivity) remains the same.

MUSCARINE comes from *Amanita muscaria*, a poisonous native red fungus with white spots. Muscarine is hallucinogenic and it competes with acetylcholine for certain neurological receptors.

H & P section 1.3 (p 8) and end-papers

Question 14

(a) Draw all the isomers of molecular formula C_4H_8.
(b) Use these isomers to explain two types of isomerism.

Answer

(a) There are six isomers: five structural and a stereoisomer of one of these:

A

B

C

D

E

F

Structural isomers *can* have different functional groups; stereoisomers must have the same functional groups.

H & P section 1.5 (p 12)

(b) (i) *Structural isomerism*: same atoms but connected in a different order, e.g. **A**, **B** (or **C**), **D**, **E** and **F**. These all have different chemical and physical properties.

(ii) *Stereoisomerism*: same atoms, same order, but a different arrangement in space, e.g. **B** and **C**. **B** (*cis*) and **C** (*trans*) differ because of the high-energy barrier to rotation about the central C=C, so these are *cis/trans* (or geometric) stereoisomers.

Question 15

Draw a dot diagram, and think of the reason why compounds show geometric (*cis/trans*) isomerism.

Explain how $CH_3CH=NH$ shows geometric isomerism.

....

Answer

The only possible source of geometric (*cis/trans*) isomerism is the double bond, C=N. Rotation about a double bond is difficult (needs high energy) because the π bond must be broken. The two H atoms on C=N can be either on the same side or on opposite sides.

H & P section 1.5 (p 13)

The nonbonded pair counts as another 'group' attached to N.

Question 16

Making molecular models will help.

Draw the structures of any possible *cis/trans* (geometric) isomers of the following molecules.

SORBIC ACID and its K^+ salt are used as food preservatives, for example in pastry.

(a) $CF_3CH_2O(CH_3)C=CH_2$ Fluorexene

(b) $CH_3CH=CHCH=CHCOOH$ Sorbic acid

(c) β-Ionone

....

Answer

FLUOREXENE is an anaesthetic. It has two identical (H) substituents on the right-hand end of the C=C, so has no *cis/trans* isomers.

(a) Fluorexene

no *cis/trans* isomers

(b) Sorbic acid

trans, trans   *trans, cis*   *cis, trans*   *cis, cis*

(c) β-Ionone

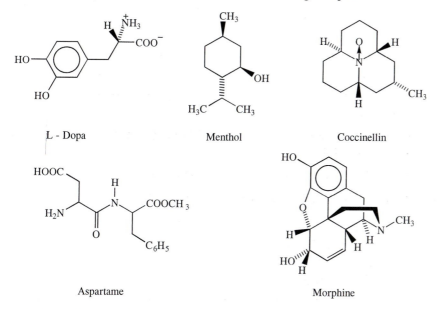

There are no other *cis/trans* isomers because the second C=C is in a six-membered ring. Cyclohexene cannot accommodate a *trans* C=C.

cis only   *trans* not stable: too strained

β-IONONE is used in perfumery. At high concentrations β-ionone has a woody smell; when dilute it gives a violet scent. It is one of the most powerful odorants known: its detection threshold in water is 0.007 parts per billion.

H & P sections 1.3, 1.5 (pp 7, 3)

Question 17

Pick out the asymmetric carbon atoms in the following compounds.

L - Dopa

Menthol

Coccinellin

Aspartame

Morphine

An asymmetric C atom bears four different substituents. It helps if you put 'missing' H atoms into skeletal (stick) diagrams.

L-DOPA plays a rôle in central nervous system transmission. Daily doses of L–dopa have been used to help those suffering from Parkinson's disease.
For menthol, see Question 13.
For coccinellin, see Question 9.
For aspartame, see Question 12.

Answer

The asymmetric carbon atoms are marked with asterisks.

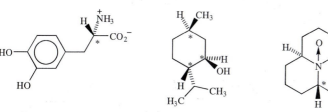

L - Dopa : 1 *   Menthol : 3 *   Coccinellin : 2 *

The [structure] of coccinellin is **not** asymmetric: the two other substituents are identical (follow them round).

MORPHINE is a potent natural analgesic, but it is addictive and has other undesirable side-effects.
The poppy produces only one stereo-isomer of morphine.

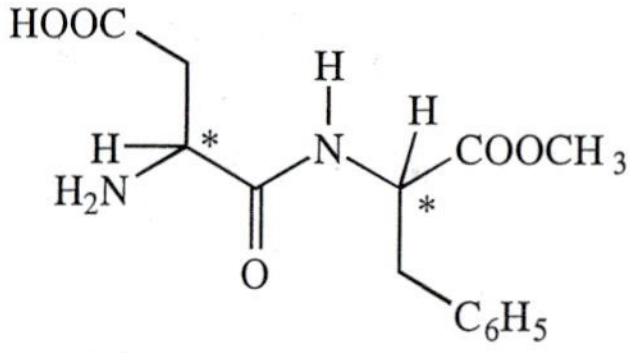

H & P section 1.5 (p 14)

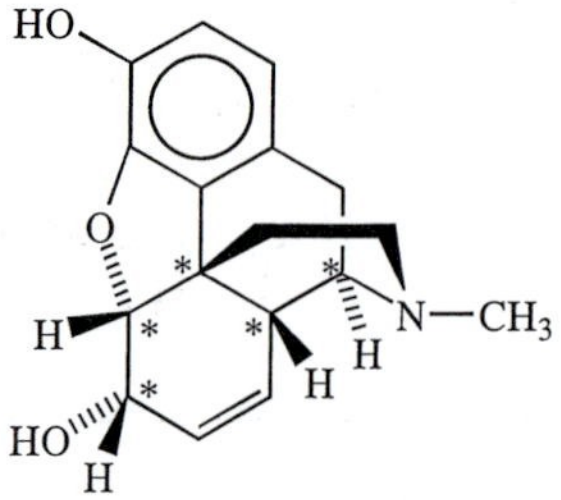

Aspartame: 2*

Morphine: 5*

Question 18

Draw diagrams to show:

(a)   the other stereoisomer of L-dopa, **A**.

Look for the asymmetric carbon atoms.

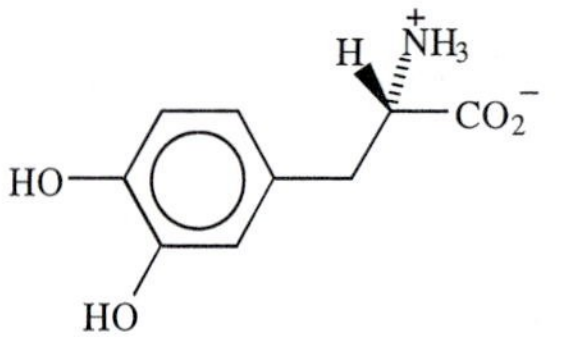

(b)   both stereoisomers of carvone, **B**.

(c)   the non-superimposable mirror image isomer of nootkatone, **C**.

A L-Dopa

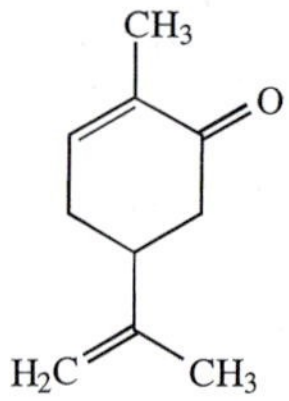

B carvone

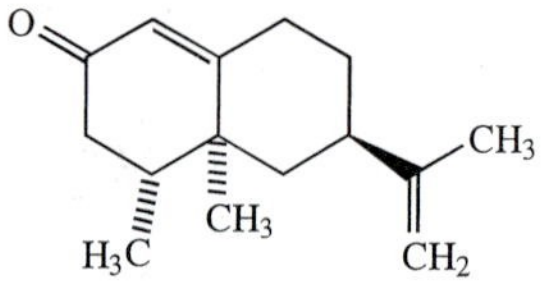

C nootkatone

....

Answer

(a)   **A**

It is easier to check these if you draw the two isomers to show their mirror image relationship. For the biological rôle of L-dopa, see Question 17.

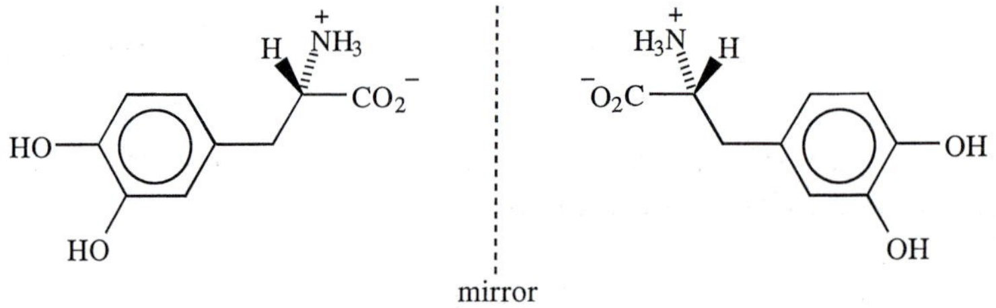

mirror

(b)   **B**

The two CARVONES smell very different: (+) carvone (on the right) smells of caraway seeds and (−) carvone (on the left) smells of spearmint. The (+) and (−) show that the compound rotates the plane of plane polarised light to the right (+) or to the left (−).

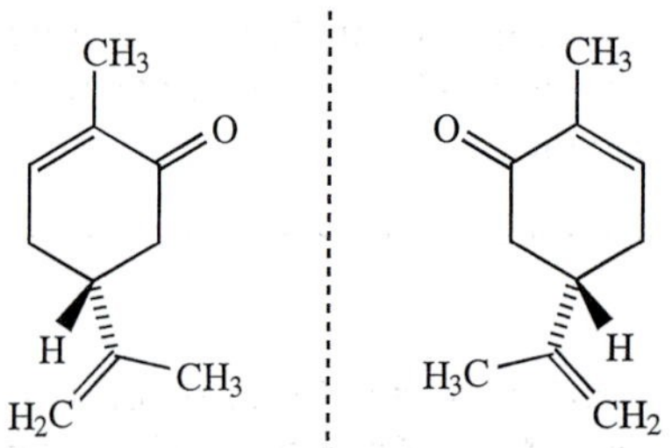

(−) Carvone

(+) Carvone

which can also be drawn as

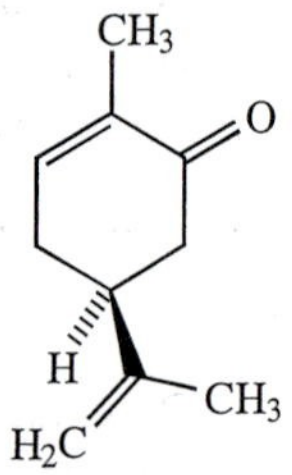

This is the same as the second structure but drawn back to front

(c) **C**

NOOTKATONE, **C**, gives grapefruit its characteristic smell and also contributes to its bitter flavour. Its non-superimposable mirror image isomer has a woody, spicy smell.

H & P section 1.5 (p 14)

Question 19

(a) Work out the shape of allene, $H_2C{=}C{=}CH_2$.

(b) Draw the two stereoisomers of $CH_3CH{=}C{=}CHCH_3$.

To get at the arrangement of π bonds around the central carbon, think of the carbons of ethyne.

Answer

(a)

$$\underset{H}{\overset{H}{\diagdown}}C{=}C{=}C\overset{\text{\tiny''H}}{\underset{H}{\blacktriangleleft}}$$

The central carbon makes two σ bonds and two π bonds, just like ethyne's carbon atoms.

Ethyne: central C:

The allene C–C–C angle about the central sp hybridised carbon is 180°, the same as C–C–H in ethyne.

Now add an ethene-like CH_2 on each σ bond of the central C, oriented so that they can combine to make the two mutually perpendicular π bonds.

This is difficult to visualise: it is easier to see with a model.

which is

$$\underset{H}{\overset{H}{\diagup}}C{=}C{=}C\overset{\text{\tiny''H}}{\underset{H}{\blacktriangleleft}}$$

The plane occupied by the CH_2 at one end is perpendicular to the plane occupied by the CH_2 at the other end.

(b)

Two non-superimposable mirror image isomers. The molecules have no plane or centre of symmetry, and each will rotate the plane of plane polarised light to the same extent, but in opposite directions.

H & P sections 1.3, 1.5 (pp 3, 4)

Question 20

(a) (i) Draw diagrams to show all seven stable structural isomers of molecular formula C_3H_6O (omit any with C=C–OH which will not be stable enough for isolation). State the functional group(s) in each structure.

See end papers for functional groups.

(ii)   Draw diagrams to illustrate any stereoisomerism shown by these structural isomers.

(b)   Draw a structural isomer of C_4H_8O which is a secondary alcohol and shows *two types of stereoisomerism*. Explain your answer.

Answer

(i)

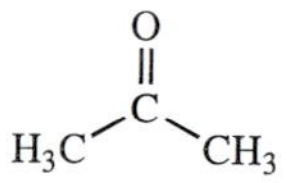

ketone 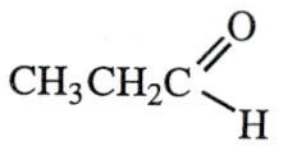   aldehyde

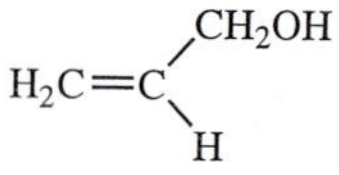

alkene and   alkene and
primary alcohol   ether

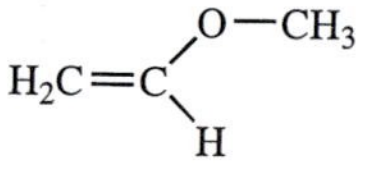

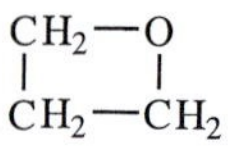

ether 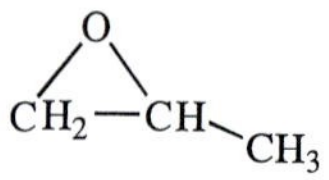   ether 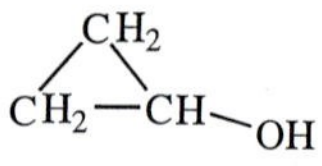   secondary alcohol

(ii)   Geometric (*cis/trans*) isomers: none.
Optical isomers: the 3-ring ether only.

One carbon atom has four different substituents: H, O, CH_2 and CH_3.

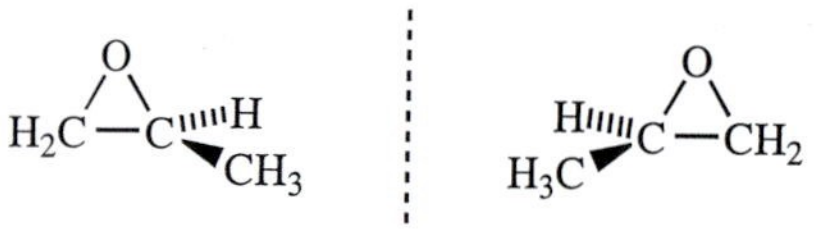

(b)

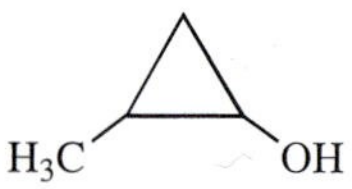

Its two geometric isomers are:

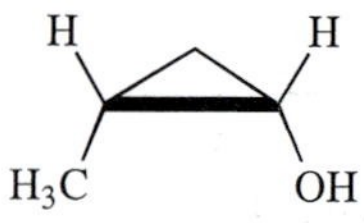

                          and

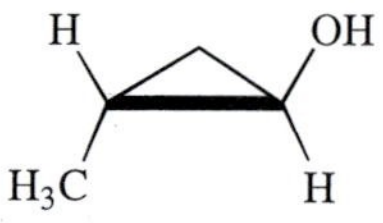

cis   *trans*

and each has a non-superimposable mirror image isomer, e.g.

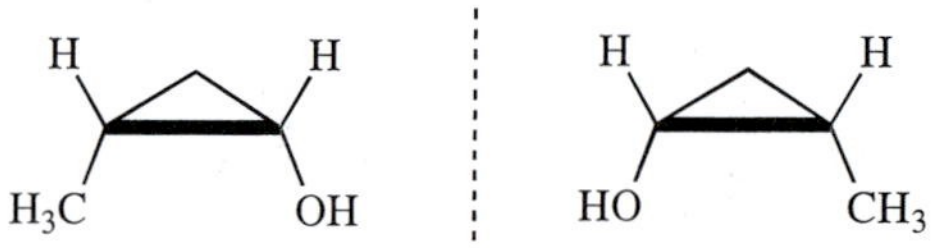

Here are the two *trans* optical isomers.

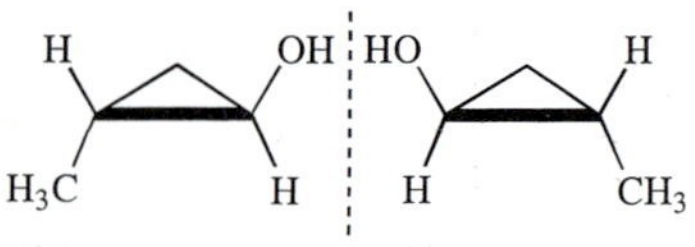

(both are *cis*)

H & P sections 1.3, 1.5 (pp 8, 12)

So this structural isomer has four stereoisomers, all of which rotate the plane of plane polarised light.

Question 21

Use δ^+ and δ^- in diagrams to identify the largest bond dipoles in each of the following molecules. Then decide whether or not each molecule has a permanent dipole moment.

Draw full 3D structures of these molecules. Look up *electronegativity*. C–H dipoles are small enough to be ignored in this question.

(a) CH_3Cl and CH_2Cl_2

(b) CO_2 and CH_3COCH_3

(c) All the isomers of $C_2H_2F_2$.

Answer

(a)

Structures		
	H$_3$C$_{\delta+}$—Cl$^{\delta-}$	H$_2$C$_{2\delta+}$—Cl$^{\delta-}$ (Cl$^{\delta-}$ below)
Dipole moment?	Yes	Yes
Bond dipole vectors	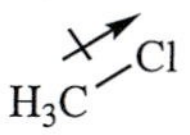 H$_3$C→Cl	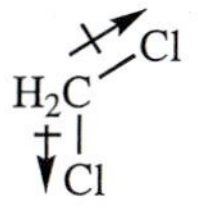 H$_2$C→Cl (Cl below)
Do these vectors cancel out?	No	No

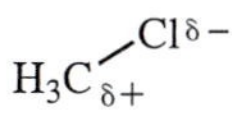

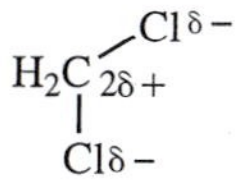

A dipole δ^+–δ^- is a vector and can be represented by the symbol $\longleftrightarrow$, where the arrowhead is the δ^- end of the dipole.

It is a common error to draw structures with incorrect 90° bond angles, for example as shown below.

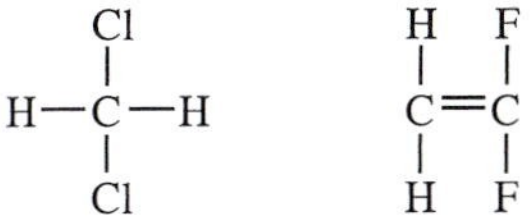

Symmetry is important here as the bond dipole vectors may cancel out: see CO_2 in (b) and *trans* CHFCHF in (C). In CH_2Cl_2 the bond dipole vectors are at ~109° to each other and do *not* cancel out. Hence CH_2Cl_2 has a permanent dipole moment.

(b)

Structures		
	δ^- $2\delta+$ δ^- O=C=O	O$^{\delta-}$ C$^{\delta+}$ H$_3$C—CH$_3$
Dipole moment?	No	Yes
Bond dipole vectors	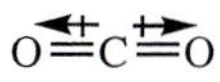 O⇇C⇉O	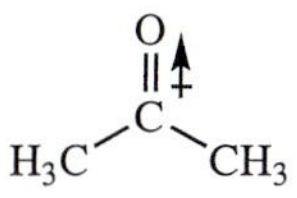 H$_3$C—C(=O↑)—CH$_3$
Do these vectors cancel out?	Yes	No

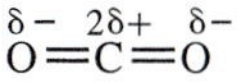

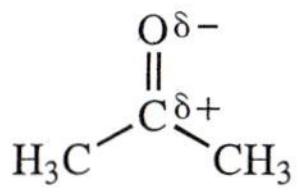

(c)

Structures	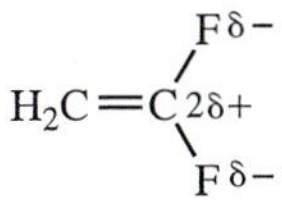	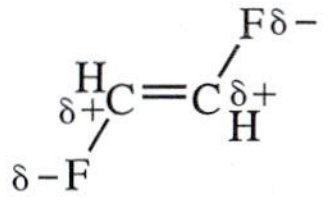	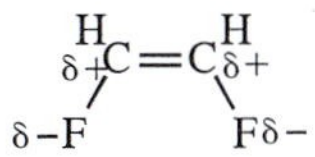
	H$_2$C=C$_{2\delta+}$ (F$^{\delta-}$, F$^{\delta-}$)	$_{\delta+}$H C=C $_{\delta+}$H (F$^{\delta-}$ top, $_{\delta-}$F bottom)	H C=C H ($_{\delta+}$) ($_{\delta-}$F, F$^{\delta-}$)
Dipole moment?	Yes	No	Yes
Bond dipole vectors	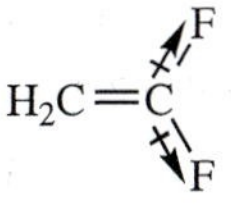 H$_2$C=C (F, F)	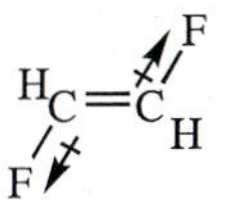 H C=C H (F, F)	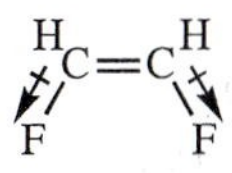 H C=C H (F, F)
Do these vectors cancel out?	No	Yes	No

The first compound is a *structural isomer* of the second and third, which are *geometric* isomers of each other.

H & P section 1.6 (p 15)

Question 22

Using diagrams, explain why propanone

(a) has a higher boiling point than 2-methylpropene;

(b) is more soluble in water than 2-methylpropene.

....

Analyse liquid propanone, liquid 2-methyl propene and mixtures of each with water for all intermolecular interactions.

Answer

(a) There are higher intermolecular forces in propanone, due to dipole–dipole interactions which are stronger than the van der Waals forces also present in both liquids. These intermolecular forces hinder molecules from escaping from the liquid. There is little polarisation in the alkene.

The boiling of a liquid does *not* involve the breaking of *intramolecular* bonds within the molecule; it involves the breaking of bonds, usually weak ones, between molecules (*intermolecular* bonds or forces).
Water is *not* involved in the boiling of any pure liquid (except water itself). The ability of a compound to form hydrogen bonds *to water* is not relevant in part (a).

M_r = relative molecular mass

Possible forces	Propanone ($M_r = 58$)	Propene ($M_r = 56$)
van der Waals	✓ weak	✓ weak
dipole–dipole	✓ stronger	x
H bonding	x	x
Ionic	x	x

(b) Propanone can form hydrogen bonds with water molecules

Hydrogen bonds are formed between a hydrogen atom bonded to a highly electronegative atom (such as O, N and F) and a second highly electronegative atom.
2-Methylpropene has *no* highly electronegative atoms.

which are comparable in strength to those between water molecules, so the two liquids mix. 2-Methylpropene does not form hydrogen bonds with water; it is energetically favourable to preserve the water–water hydrogen bonding, so the two liquids remain separate.

H & P sections 1.6, 1.7 (pp 15, 17)

Question 23

Draw a diagram to show the main ways in which paracetamol could be involved in hydrogen bonding with water molecules.

PARACETAMOL is a widely used painkiller.

Paracetamol

Answer

Strong hydrogen bonds are formed between H covalently bonded to O, F and N and a second electronegative atom (O, F or N). The higher the electron density on the second atom, the better.

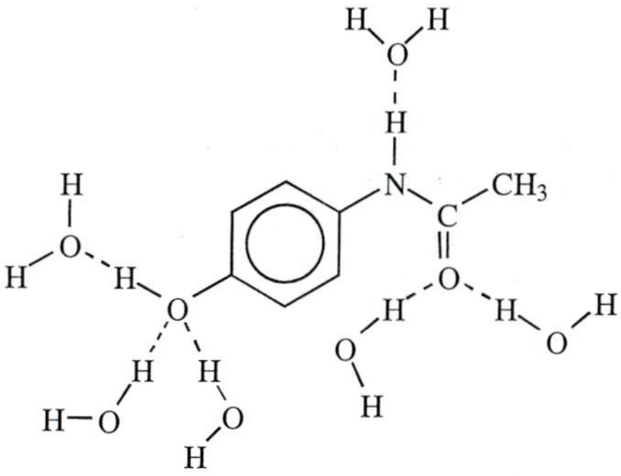

[6 sites]

Remember that the amide nitrogen nonbonded pair is delocalised, reducing the electron density on N and increasing the electron density on O.

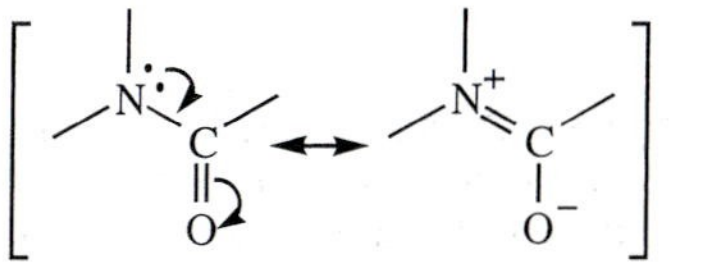

This means that hydrogen bonding to the O of an amide is much stronger than to the N.

H & P sections 1.6, 1.7 (pp 15, 17)

Question 24

Explain why 'soluble' aspirin is likely to be more soluble in water than regular aspirin.

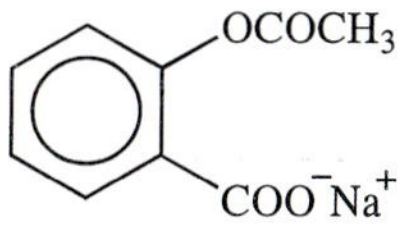

'Soluble' aspirin   Aspirin

What kind of intermolecular attractions with water would be operating in each case?

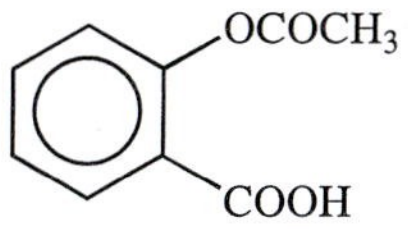

Answer

Both compounds can form hydrogen bonds extensively with water.

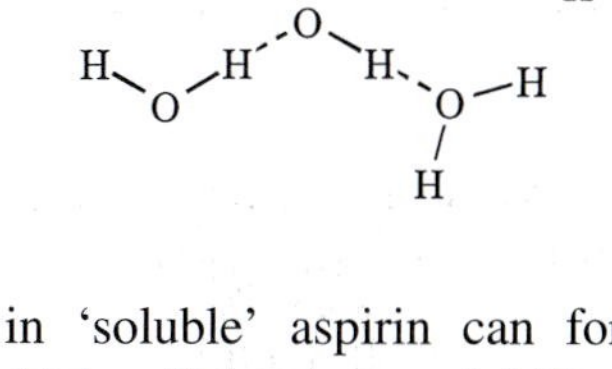

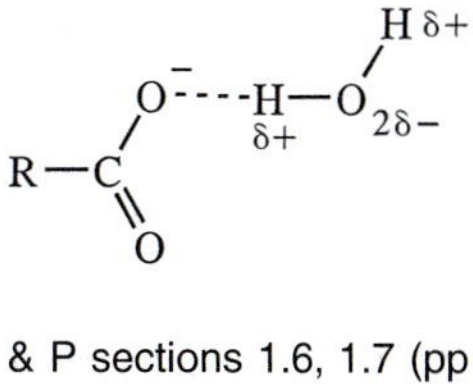

In addition, both ions in 'soluble' aspirin can form stronger ion–dipole interactions with water which will improve solubility.

H & P sections 1.6, 1.7 (pp 15, 17)

Question 25

Glucose is converted into hydroxymethylfurfural (HMF) during browning reactions which accompany cooking.

Glucose

Would you expect HMF to be soluble in water? Explain your answer fully with diagrams.

Answer

HMF has no less than four different sites for hydrogen bonding to water in a relatively small molecule, and is therefore likely to be soluble in water.

2 Mechanisms

2.2 Worked examples

Question 1

Classify each of the following reactions as substitution, addition or elimination.

(a) $CH_3CH_2CH_2Br$ + KOH $\longrightarrow$ $CH_3CH=CH_2$ + KBr + H_2O

Add up the molecular formulae of the organic molecules on each side of the equation, and look at the difference. Molecular formulae should be written with C, H first, then other atoms in alphabetical order.

(b) $\longrightarrow$ + H_2O

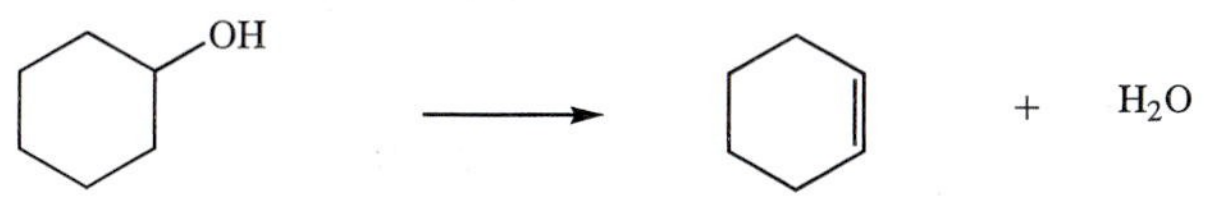

This reaction is usually acid-catalysed.

(c)   + KOH $\longrightarrow$ + KBr

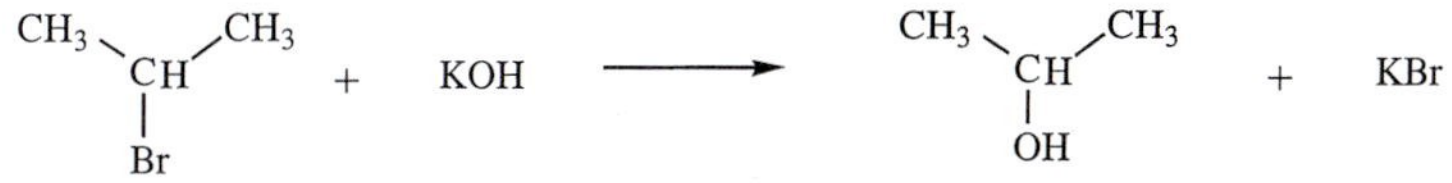

(d) CH_3CHO + HCN $\longrightarrow$

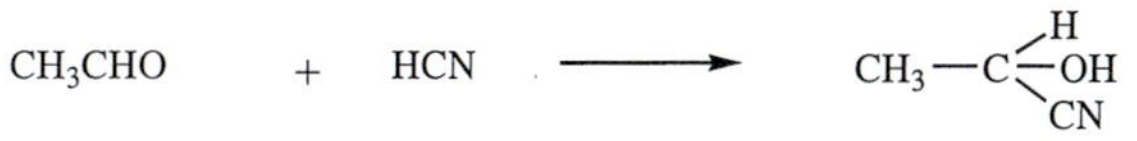

Cyanide ions are needed to make this reaction work, so KCN (or a base to deprotonate HCN) is added. More commonly the HCN is made in the reaction mixture from KCN and H_2SO_4.

(e)   + Br_2 $\longrightarrow$

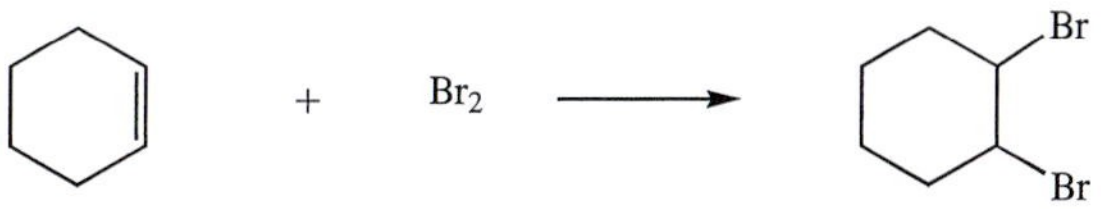

(f) CH_3COCl + H_2O $\longrightarrow$ CH_3COOH + HCl

Answer

(a)   Elimination: HBr has been lost from C_3H_7Br.

(b)   Elimination: H_2O has been lost.

(c)   Substitution: OH has replaced Br.

(d)   Addition: H and CN (HCN) have been added across the C=O double bond.

(e)   Addition: Br_2 added across the C=C bond.

(f)   An overall substitution: OH replaced Cl. This can also be classified as addition–elimination as this is how the reaction occurs (the mechanism).

$C_3H_7Br \xrightarrow{-HBr} C_3H_6$

$C_3H_7Br \xrightarrow{-Br, +OH} C_3H_8O$

$C_2H_3ClO \xrightarrow{-Cl, +OH} C_2H_4O_2$

H & P Section 2.1 (p 18)

Question 2

Classify each of the following as a nucleophile, an electrophile or a radical:
(a) a chloride ion;   (b) a proton;   (c) a chlorine atom;
(d) Br^+;   (e) HBr.

Dot diagrams will be helpful.

....

Answer

(a) $\left[:\!\overset{..}{\underset{..}{Cl}}\!: \right]^-$ *Nucleophile*: four nonbonded pairs of electrons. It can donate an electron pair to make a new bond.

(b) $[\,H\,]^+$ *Electrophile*: can accept a pair of electrons to make a new bond, filling the 1s shell.

Needs one more electron to complete the shell.

(c) $:\!\overset{..}{Cl}\!\cdot$ *Radical*: one unpaired electron.

Loss of one electron from a bromine atom (seven electrons) gives the six-electron, positively charged Br^+ ion.

(d) $\left[:\!\overset{..}{\underset{..}{Br}} \right]^+$ *Electrophile*: can accept a pair of electrons, making eight altogether in the outer shell.

(e) $H\!:\!\overset{..}{\underset{..}{Br}}\!:$ or $\overset{\delta+}{H}\!\!-\!\!\overset{\delta-}{Br}$, a polarised molecule and an *electrophile*. It can accept a pair of electrons to form a new bond to H at the same time as the H–Br bonding pair goes off with the bromine atom to form the stable bromide ion; e.g. with the nucleophile $C_6H_5NH_2$ (watch the non-bonded $*$ pair):

H & P sections 2.2, 2.3 (pp 19, 20)

$$C_6H_5\!:\!\overset{H}{\underset{H}{N}}\!\overset{*}{_*} \quad + \quad H\!:\!\overset{..}{\underset{..}{Br}}\!: \quad \longrightarrow \quad \left[C_6H_5\!:\!\overset{H}{\underset{H}{N}}\!\overset{*}{_*}H \right]^+ \left[:\!\overset{..}{\underset{..}{Br}}\!: \right]^-$$

Question 3

Curly arrows show *pairs* of electrons moving, so they must begin from a pair of electrons, either nonbonded or bonded. They must end on an atom or in a bond.

Fish-hook arrows show the movement of *single* electrons.

Translate each of the following dot diagram equations into the curly arrow version, using structural formulae.

(a)

then

(b)

(c)

Answer

(a)

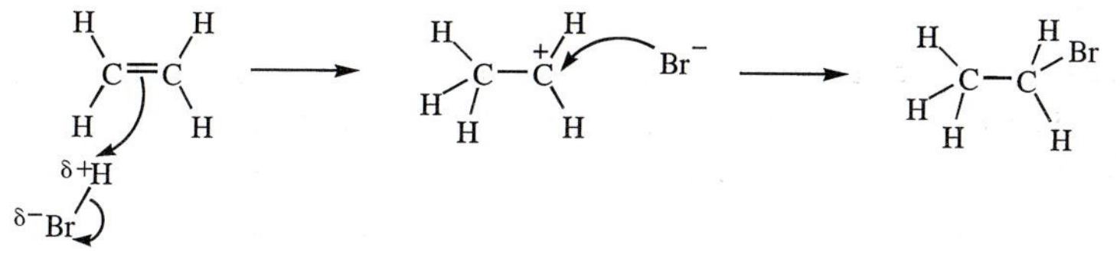

The nucleophiles here are the C=C (π electrons) and Br^-; the electrophiles are the δ^+H of HBr and the $^+CH_2$ of $C_2H_5^+$.

(b)

$$NC \overset{..}{:} \quad H_3C \overset{\frown}{-}Br \quad \longrightarrow \quad NC-CH_3 \quad + \quad Br^-$$

(c)

$$Br\overset{\frown}{-}Br \quad \longrightarrow \quad 2Br^\bullet$$

$$H_3C\overset{\frown}{-}H \quad \overset{\frown}{} Br \quad \longrightarrow \quad H_3C\bullet \quad + \quad H-Br$$

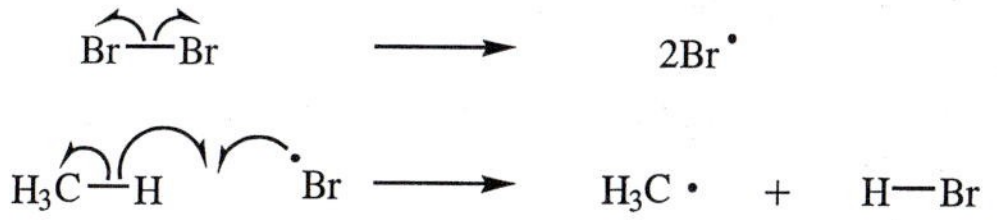

This is *homolytic* bond breakage in which the electrons of the bonding pair go to different atoms.

H & P section 2.3 (p 20)

Question 4

(a) Classify the following reaction as homolytic or heterolytic, justifying your answer by using a dot diagram.

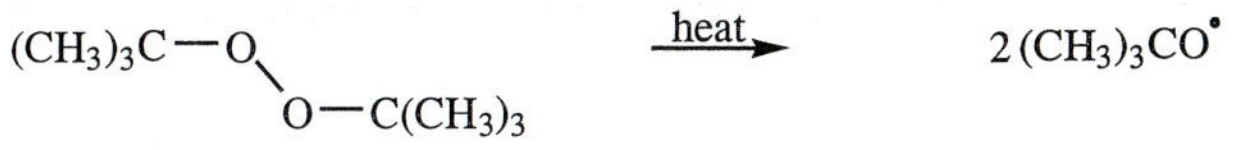

$$(CH_3)_3C-O \overset{O-C(CH_3)_3}{} \quad \xrightarrow{\text{heat}} \quad 2\,(CH_3)_3CO^\bullet$$

(b) Give an example of a homolytic reaction involving chlorine, with equations to show the stages of initiation, propagation and termination. Overall, is your reaction an addition, substitution or elimination reaction?

(c) Explain in words what is happening in the following mechanism.

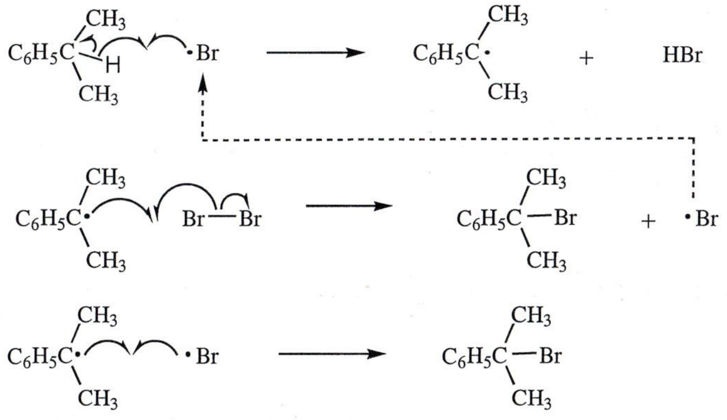

Remember that a fish-hook arrow ⌢ shows the movement of *one* electron (a full curly arrow ⌢ shows the movement of a *pair* of electrons).

(i) $$Br_2 \quad \xrightarrow{\text{irradiate}} \quad 2Br^\bullet$$

(iia)

$$C_6H_5C \quad \longrightarrow \quad C_6H_5C^\bullet \quad + \quad HBr$$

(iib)

$$C_6H_5C^\bullet \quad \longrightarrow \quad C_6H_5C-Br \quad + \quad \bullet Br$$

(iii)

$$C_6H_5C \quad \bullet Br \quad \longrightarrow \quad C_6H_5C-Br$$

Answer

(a) Homolytic: the two electrons of the O–O bond are unpaired and go to two different atoms. Two new radicals are created, each with one unpaired electron.

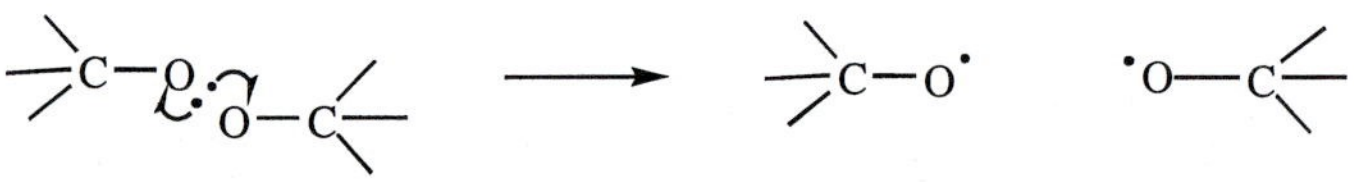

(b) For example, the chlorination of methane.

Overall $CH_4 + Cl_2 \xrightarrow{irradiate} CH_3Cl + HCl$.

An H of CH_4 has been *substituted* by Cl in a homolytic mechanism involving radicals.

INITIATION $\quad Cl\!-\!Cl \longrightarrow 2Cl^\bullet$

or

PROPAGATION $\quad CH_4 + Cl^\bullet \longrightarrow {}^\bullet CH_3 + HCl$

$\qquad\qquad\qquad {}^\bullet CH_3 + Cl_2 \longrightarrow CH_3Cl + Cl^\bullet$

TERMINATION $\qquad 2\,Cl^\bullet \longrightarrow Cl_2$

or $\quad 2\,{}^\bullet CH_3 \longrightarrow CH_3CH_3$

or $\quad Cl^\bullet + {}^\bullet CH_3 \longrightarrow CH_3Cl$

(c) (i) Homolytic fission of the Br–Br bond to form two radicals—initiation, as in part (b). One electron goes to each bromine atom.

(iia) Homolytic fission of the C–H bond induced by the bromine radical to give HBr and a carbon-centred radical.

(iib) The carbon radical reacts with bromine to form the new C–Br bond and to regenerate a bromine radical.

The Br$^\bullet$ then goes back to repeat (iia).

(iia) and (iib) are the propagation steps, and both involve homolytic bond fission.

(iii) A carbon radical and a bromine radical combine their single electrons to make a covalent C–Br bond. This is an example of a termination reaction; two radicals involved in the propagation steps are destroyed, so two potential chains are stopped.

These explanations are very long-winded: you can see why chemists prefer to use curly arrows to describe mechanisms!

Alternative termination stages would be

$$2\dot{Br} \longrightarrow Br_2$$

or

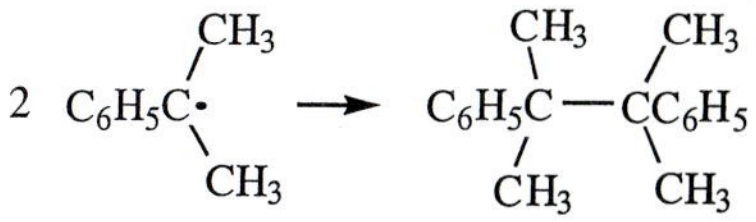

That is, any combination of two radicals taking part in the propagation steps (here, iia and iib).

H & P sections 2.1, 2.2, 2.3 (pp 18, 19, 20)

Question 5

(a) Draw a dot diagram for CN^-, and explain why the carbon bears a negative charge.

(b) Give an example of the use of cyanide ion in a nucleophilic substitution reaction. Using your example, explain the meaning of the terms *nucleophile* and *heterolytic*.

(c) Explain in words what is happening in the following mechanism:

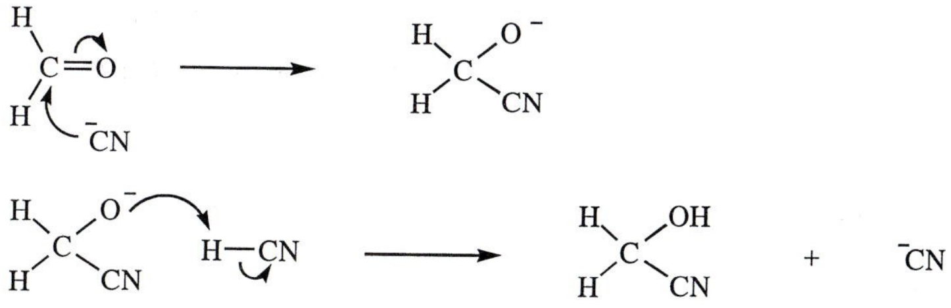

....

Answer

(a) $[:C \vdots\vdots N:]^-$. There is a triple bond between C and N, and each atom has a share in eight electrons. Dividing the bonding pairs evenly between C and N, the N has 5 electrons (neutral N atom) and the C has 5 which is one more than a neutral carbon atom, so the C is negatively charged.

(b) For example, reaction of CN^- with a halogenoalkane to give a nitrile:

$$C_2H_5Br + CN^- \rightarrow C_2H_5CN + Br^-$$

Mechanism:

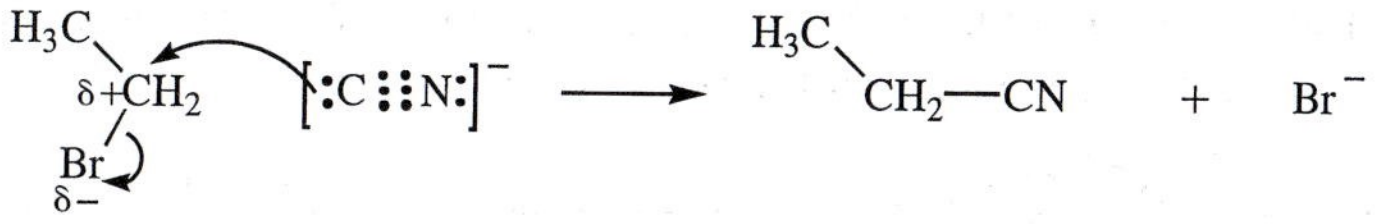

CN^- acts as a *nucleophile* because it uses a nonbonded pair of electrons on carbon to form the new bond. All electrons remain paired so the bond-making and bond-breaking are *heterolytic*, the pair of bonding electrons coming from (or going to) the same atom.

(c)   A nonbonded pair on the carbon of CN^- (the nucleophile) is used to make a new bond to the carbonyl carbon as the π pair of the C=O bond becomes a nonbonded pair on the electronegative oxygen, giving the O a negative charge. Next a nonbonded pair on the $-O^-$ forms a bond to a hydrogen of HCN as the H–CN bond breaks to re-form CN^-. All reactions are *heterolytic* (paired electrons).

Overall, HCN is consumed in this reaction.

H & P sections 2.1, 2.2, 2.3 (pp 18, 19, 20)

Question 6

(a)   Classify the following reaction as homolytic or heterolytic, using dot diagrams to justify your answer.

$$(CH_3)_3CBr \longrightarrow (CH_3)_3\overset{+}{C} \quad \overset{-}{Br}$$

(b)   Explain in words what is shown by the following mechanism.

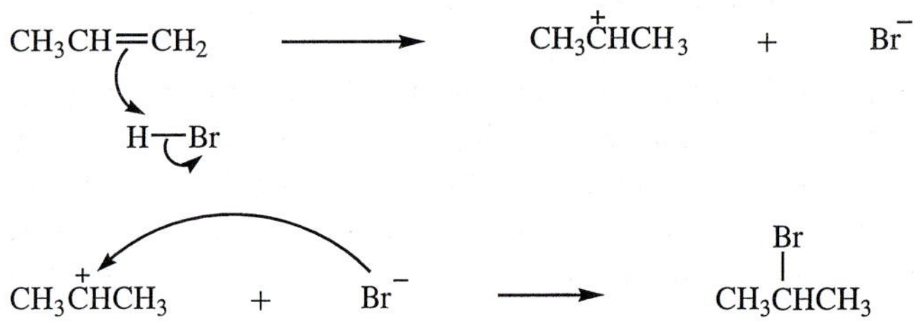

$$CH_3CH{=}CH_2 \longrightarrow CH_3\overset{+}{C}HCH_3 \quad + \quad \overset{-}{Br}$$

H—Br

$$CH_3\overset{+}{C}HCH_3 \quad + \quad \overset{-}{Br} \longrightarrow \overset{\displaystyle Br}{\underset{\displaystyle CH_3CHCH_3}{|}}$$

Remember that there are nonbonded pairs on Br^-.

....

Answer

(a)

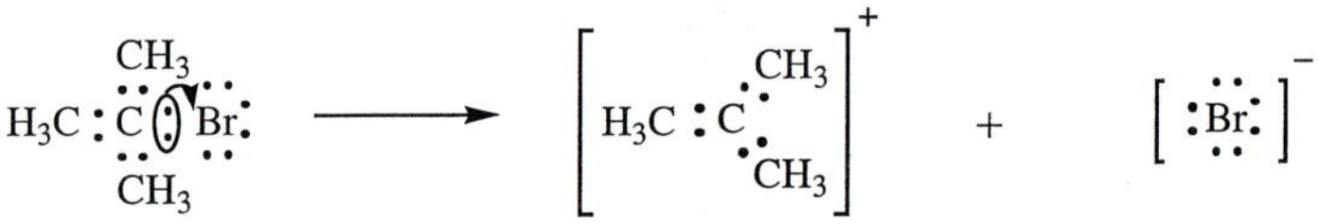

$$\underset{\displaystyle CH_3}{\overset{\displaystyle CH_3}{H_3C\!:\!C\!:\!Br\!:}} \longrightarrow \left[\underset{\displaystyle CH_3}{\overset{\displaystyle CH_3}{H_3C\!:\!C}} \right]^+ \quad + \quad \left[:\!Br\!: \right]^-$$

Heterolysis: both C–Br bonding electrons leave with the Br, giving Br a negative charge and the carbon a positive charge.

How can you work out the charge on carbon from the dot diagram? The central C has six outer shell electrons, in three bonding pairs. If one of each bonding pair is supplied by the methyl group carbon, then the central carbon only provides three electrons. This is one less than in a neutral carbon atom, so it is positively charged.

Alternatively, if we start with a neutral carbon atom and three methyl radicals, we make a neutral radical.

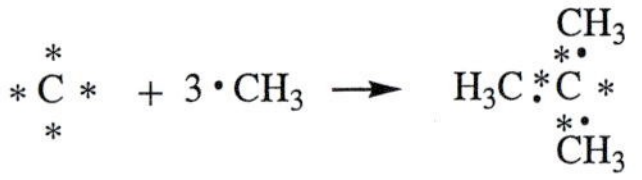

$$*\overset{*}{\underset{*}{C}}* \ + \ 3\cdot CH_3 \longrightarrow \underset{\displaystyle \overset{*}{C}H_3}{\overset{\displaystyle CH_3}{H_3C\!\overset{*}{*}\!C\!*}}$$

We then need to take one electron away from this, giving it a positive charge.

Br^- has four nonbonded pairs.

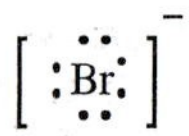

$$\left[:\!\overset{\displaystyle ..}{\underset{\displaystyle ..}{Br}}\!: \right]^-$$

H & P sections 2.1, 2.2, 2.3 (pp 18, 19, 20)

(b)   The π pair of the propene double bond (acting as a nucleophile, electron-rich) forms a new bond with the electron-deficient δ^+ hydrogen of HBr as the H–Br bonding pair leaves with the bromine as Br^-. A carbocation intermediate is formed.

The Br^- uses a nonbonded pair to make a bond to the electron-deficient carbon of the $CH_3CH_2^+$, forming the new covalent bond.

All reactions are heterolytic, involving the electrons in the broken bond remaining paired and both going to one atom.

Question 7

An energy profile for the nucleophilic substitution reaction

$$I^- + CH_3CH_2Br \rightarrow Br^- + CH_3CH_2I$$

is sketched below.

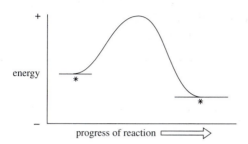

Redraw the diagram for yourself.
(a) Label your diagram to show the species at the asterisks.
(b) Draw in the activation energy E_{af} for the forward reaction.
(c) The reaction is reversible. Draw in the activation energy E_{ab} for the back reaction.
(d) Put 'TS' in the correct place in the diagram for the transition state.
(e) Suggest a mechanism for this reaction, explaining why it is called a *nucleophilic* substitution reaction. What is acting as the *electrophile*?
(f) Classify the reaction as *homolytic* or *heterolytic*, giving your reasoning.
(g) Predict a structure for the transition state, TS, for this reaction.

....

Answer

The reaction proceeds from left to right by convention.

(a), (b), (c) and (d) answers are all shown on the same diagram:
(b), (c) This E_{af} is the smallest amount of energy needed to go over the energy hump from starting materials to products. E_{ab} is the lowest energy needed to go backwards, from $C_2H_5I + Br^-$ to $C_2H_5Br + I^-$.
(d) The transition state is the species at the top of the energy hump.
(e) Mechanism

$$I \cdots CH_2 - Br \longrightarrow I - CH_2 + Br^-$$
$$\underset{CH_3}{|} \qquad\qquad \underset{CH_3}{|}$$

I^- is the nucleophile, sharing a nonbonded pair of electrons with the carbon to form the new bond as Br departs with its bonding pair to form Br^-.

This is a bimolecular nucleophilic substitution reaction, abbreviated to S_N2. Rate $\propto [C_2H_5Br][I^-]$

Nucleophiles react with electrophiles, so the carbon of the $-CH_2Br$ group acts as the electrophile.

(f)   TS is $\overset{\delta-}{I}$---CH_2---$\overset{\delta-}{Br}$. Here is a way to work it out.
$\quad\quad\quad\quad\quad\quad\quad\quad\quad$ |
$\quad\quad\quad\quad\quad\quad\quad\quad\quad CH_3$

(1)   *Bonds*   The CH_3CH_2 remains unchanged, so draw that in. C–I is made, C–Br broken, so both will be partially formed in the TS

$$\left(\begin{array}{c} I\text{---}CH_2\text{---}Br \quad \text{so far} \\ | \\ CH_3 \end{array} \right)$$

(2)   *Charges*   A transition state is in between starting materials and products, so the charge on any atom in the TS will probably be between that borne by that atom in the starting material and in the product. The atoms in CH_3CH_2Br and CH_3CH_2I bear no full charges.

Consider the *iodine* atom. It starts off negatively charged (–1) in iodide ion and will have no charge (0) in CH_3CH_2I. In between it must be partially negative, that is δ^-.

In the same way, the *bromine* atom goes from no charge (0) to negative (–1) so will also be δ^- in the TS.

Carbon bears no formal charge throughout so is uncharged.

So the full TS is $\overset{\delta-}{I}$---CH_2---$\overset{\delta-}{Br}$
$\quad\quad\quad\quad\quad\quad\quad\quad\quad\quad$ |
$\quad\quad\quad\quad\quad\quad\quad\quad\quad\quad CH_3$

Question 8

Predict the shape of the energy profile for the reaction of radioactive iodide ion $^{131}I^-$ with iodoethane, labelling the maxima and minima with diagrams of the species concerned. Estimate the equilibrium constant for this reaction.

....

Answer

Writing ^{131}I as $^*I$

$$^*I^- \quad + \quad C_2H_5I \quad \rightleftharpoons \quad C_2H_5I^* \quad + \quad I^-$$

The products are the same as the starting materials, and have the same energy. C–I bonds are made and broken, iodine changes from –1 to 0 or 0 to –1, so in TS is δ^-. This δ^- is $\frac{1}{2}-$ because the TS is symmetrical. No charge on C.

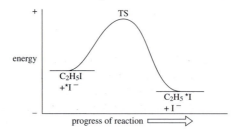

Energy change $= -RT \ln K$
$\quad$ but the energy change is zero
Therefore $O = RT \ln K$ therefore $K = 1$.

Side notes (left margin):

We do not know how big $\delta-$ is: probably *not* $\frac{1}{2}-$, because the TS is not symmetrical. Complex calculations and/or more experimental data can give us an estimate of this.

H & P sections 2.1, 2.2, 2.3, 2.4 (pp 18, 19, 21, 28)

Write down the equation for the reaction first.

This is another single-stage S_N2 reaction.

H & P sections 2.3, 2.4 (pp 20, 28)

Question 9

The energy profile for the reaction

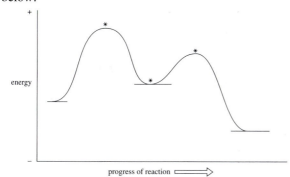

is shown below.

Suggest structures for the organic species at the starred positions and indicate the activation energy for the rate-limiting step on the energy profile, labelling it E_a.

· ·

Answer

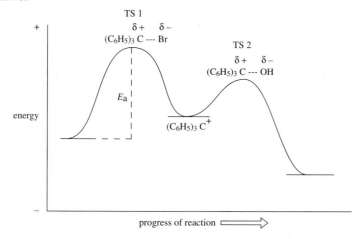

The intermediate, $(C_6H_5)_3C^+$, is at the central energy minimum. The two transition states TS1 and TS2 are at the maxima. How do you work out the structure of TS1?

(1) *Bonds* Leave alone anything unchanged across the reaction, that is, all the $(C_6H_5)_3C$ atoms. Then identify any bonds made or broken. The C–Br bond is broken, so in TS1 it will be partially broken, C ··· Br.

(2) *Charges* The charge on the central carbon atom goes from neutral (0) to +1, so in the TS it will be slightly positive, δ^+. The Br goes from 0 to –1 so it will be δ^-.

Similarly for TS2 where the C–OH bond is being formed, and the charge on C goes from +1 to 0 and on oxygen from –1 ($^-$OH) to 0.

The slower, more energy-demanding, rate-limiting step is the first one, the ionisation to the reactive carbocation intermediate.

See Question 7 for another example of how to work out TS structures.

This is a unimolecular nucleophilic substitution reaction (S_N1).
Rate $\propto [(C_6H_5)_3CBr]$.
The rate is independent of $[OH^-]$.

H & P section 2.4 (p 22)

Question 10

The sulphonation of naphthalene, **A**, can give two different products, depending on the reaction conditions used.

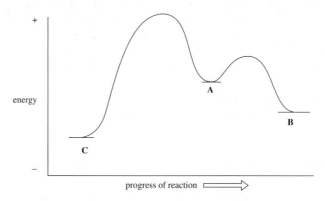

C   **A**   **B**

The energy profile for these reactions is shown below.

energy

progress of reaction ⟹

Look at the structures of **B** and **C**.

(a)   Suggest a reason why **C** is more stable than **B**.

(b)   Use the energy profile to explain
   (i)   how the major product at the lower temperature is **B**;
   (ii) how the major product at the higher temperature is **C**.

Answer

(a)   **C** is more stable than **B** because there is less crowding between C_1–H and –SO$_3$H in **C** than between C_8–H and –SO$_3$H in **B**:

This is best seen with space-filling models.

The numbering system is

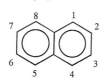

   C   **B**

(bi)   The activation energy **A** → **B** is lower than that for **A** → **C**. With less thermal energy available, not many molecules have sufficient energy to go over the **A** → **C** energy barrier. More will be able to go from **A** to **B**, so **B** is the major product.

B is referred to as the *kinetic product* (made more quickly).

(ii)   At the higher temperature, most of the molecules now have enough energy to go over from **A** to **C**. Not only that, but there is sufficient energy for **B** to go back to **A** as well. Once converted to **C**, it is still difficult to go over the highest barrier (back to **A**), so compound **C** accumulates as the major product at high temperatures.

C is referred to as the *thermodynamic product* (more stable).

H & P section 2.4 (p 29)

3 Acids and bases

3.1 What you need to read about or revise for this chapter

A Equilibria; definitions of K_a, K_b, K_w; pK_a, pK_b, pK_w, pH; and calculations involving manipulation of *all* of these.

B Factors affecting acidity and basicity of organic compounds, such as electronegativity of the atom bearing the charge; delocalisation and resonance; inductive effects.

C Effects of acidity and basicity in other areas of chemistry: indicators, interactions between molecules and ions.

3.2 Worked examples

Question 1

(a) (i) Using dot diagrams write equations to show the reactions of ammonium ion and of water to give H^+.

(ii) Write an expression for K_a for aqueous ammonium ion.

(b) The table shows the pK_a values for some acids.

Acid	pK_a
Ethanoic acid, CH_3COOH	4.8
Ammonium ion, NH_4^+	9.3
Phenylammonium ion, $C_6H_5NH_3^+$	4.6

You might find it easier to turn the pK_a values into K_a values.

$$pK_a = -\log_{10} K_a$$

Explain which is the strongest acid.

(c) Explain how ethanoic acid can also act as a base at very low pH, but ammonium ion cannot.

Bases usually have nonbonded pairs of electrons.

Answer

(a) (i)

(ii) For NH_4^+, $\quad K_a = \dfrac{[NH_{3(aq)}][H^+_{(aq)}]}{[NH^+_{4(aq)}]}$

Check this by calculating K_a values. For CH_3COOH,
$pK_a = -\log_{10} K_a$
$K_a = 10^{-pK_a}$
$\quad = 1.7 \times 10^{-5}$

Protonation on the 'carbonyl' oxygen atom gives the more stable, delocalised protonated form of the COOH group.

H & P section 3.1, 3.2 (p 33)

(b)   The strongest acid is $C_6H_5NH_3^+$.
The more highly dissociated the acid, the higher is the relative concentration of protons (H^+) at equilibrium, and the higher the value of K_a.
$pK_a = -\log_{10} K_a$,   (note the minus sign here)
Therefore: Strong acids have **high** K_a values and **low** pK_a values
Weak acids have **low** K_a values and **high** pK_a values.

(c) CH_3COOH has nonbonded pairs of electrons (on O) which can accept protons and so it acts as a weak base.

NH_4^+ has no nonbonded pairs left to bond to a proton.

Question 2

Write down the expression for K_a.

Calculate the pH of 0.01 M aqueous CH_3COOH (pK_a 4.8), showing your working.

....

Answer

If CH_3COOH is the *only* solute then $[CH_3COO^-_{(aq)}] = [H^+_{(aq)}]$.

CH$_3$COOH is a weak acid so the approximation $[CH_3COOH(aq)] = 0.01M$ is valid as long as the answer is kept to two significant figures.

H & P section 3.2 (p 33)

$$K_a = \frac{[CH_3COO^-_{(aq)}][H^+_{(aq)}]}{[CH_3COOH_{(aq)}]}$$

$$= \frac{[H^+_{(aq)}]^2}{0.01}$$

but $pK_a = -\log_{10} K_a = 4.8$
$\therefore K_a = 1.6 \times 10^{-5}$

$$1.6 \times 10^{-5} = \frac{[H^+_{(aq)}]^2}{0.01}$$

$$[H^+_{(aq)}]^2 = 1.6 \times 10^{-5} \times 0.01 = 1.6 \times 10^{-7}$$

$$pH = -\log_{10}[H^+_{(aq)}] = 3.4.$$

Question 3

(a)   An aqueous solution of 2,2-dimethylpropanoic acid (0.020 mol dm^{-3}) has pH 3.4.
  (i)   Draw the structural formula of the acid.
  (ii)   Calculate pK_a for the acid, showing your working.
(b)   Phenol has a pK_a of 9.9. Work out the pH of a solution containing $C_6H_5O^-$ and C_6H_5OH in 2:1 proportion.

....

Remember to use $\log_{10}$ and not $\log_e$; you need the log or lg button on your calculator, not ln.

The numbering of the 'propane' chain starts from the COOH carbon.

Answer

(a)   (i)

(ii)
$$K_a = \frac{[H^+_{(aq)}][A^-_{(aq)}]}{[HA_{(aq)}]}$$

$$= \frac{[H^+_{(aq)}]^2}{[HA_{(aq)}]} \quad \text{but} \quad pH = 3.4$$

$$\therefore [H^+_{(aq)}] = 3.98 \times 10^{-4} \text{ mol dm}^{-3}$$

$$= \frac{(3.98 \times 10^{-4})^2}{0.020}$$

$$\therefore K_a = 7.9 \times 10^{-6} \text{ mol dm}^{-3}$$

$$pK_a = 5.1$$

If HA is the only solute,
$$[A^-_{(aq)}] = [H^+_{(aq)}]$$

Or, more neatly, by taking logs earlier and rearranging:

$\log K_a = 2 \log [H^+_{(aq)}] - \log [HA_{(aq)}]$

$\therefore pK_a = 2 \, pH + \log [HA_{(aq)}]$

$\qquad = 6.8 + \log 0.02$

$\qquad = 5.1$

(b)
$$K_a = \frac{[C_6H_5O^-_{(aq)}][H^+_{(aq)}]}{[C_6H_5OH_{(aq)}]}$$

$$pK_a = -\log_{10}\frac{[C_6H_5O^-_{(aq)}]}{[C_6H_5OH_{(aq)}]} + pH$$

$$pH = pK_a + \log_{10}\frac{[C_6H_5O^-_{(aq)}]}{[C_6H_5OH_{(aq)}]}$$

$$= 9.9 + \log_{10}\frac{2}{1}$$

$$= 9.9 + 0.3$$

$$= 10.2$$

H & P section 3.2 (p 33)

Question 4

Compound	K_a (mol dm^{-3})
CH_3CH_2OH	1.0×10^{-16}
HCOOH	1.6×10^{-4}
$ClCH_2COOH$	1.3×10^{-3}

(a) Calculate the pK_a for each compound.
(b) Calculate the pK_b for the conjugate base of each compound, and state which is the strongest base.

To what equation does base strength refer?

Remember $pK_a + pK_b = pK_w = 14$.

....

Answer

(a) pK_a values: CH_3CH_2OH 16.0; HCOOH 3.8; $ClCH_2COOH$ 2.9.
$pK_a = -\log_{10}K_a$ so weak acids, which have low K_a values, have high pK_a values because of the negative sign.

(b) pK_b values: $CH_3CH_2O^-$ −2.0; $HCOO^-$ 10.2; $ClCH_2COO^-$ 11.1.
Use $pK_a + pK_b = pK_w = 14$.
For $HCOO^-$: $3.8 + pK_b = 14$ so $pK_b = 14 - 3.8 = 10.2$.
Ethoxide ion is the strongest base as it has the lowest pK_b and highest K_b. Strong bases have negative pK_b values, strong acids have negative pK_a values; the pK_a of HCl is −7.0.

For a base A^- in water

$$K_b = \frac{[HA_{(aq)}][OH^-_{(aq)}]}{[A^-_{(aq)}]}$$

$K_a \times K_b = [H^+_{(aq)}][OH^-_{(aq)}] = 10^{-14}$

Using the expression for K_a in the answer to 3 a ii.

H & P section 3.2 (pp 33, 34)

Question 5

It may help to use the equilibrium constant (K_a) instead of pK_a. Write out the equilibria.

Explain why the pK_a of ethanol is so different from that of (a) ethanoic acid and (b) aminoethane.

Compound	pK_a
CH_3COOH	4.8
CH_3CH_2OH	16
$CH_3CH_2NH_2$	~ 40

....

Answer

(a)   For CH_3CH_2OH and CH_3COOH, the equilibria are:

$K_a = 10^{-16}$

$$CH_3CH_2OH \rightleftharpoons CH_3CH_2O^- + H^+$$

$K_a = 10^{-4.8}$

$$CH_3COOH \rightleftharpoons CH_3COO^- + H^+$$

In each, an O–H bond is ionised, but for CH_3COO^- the resulting anion is stabilised by delocalisation (resonance).

The inductive electron-withdrawing effect of the C=O helps to weaken the adjacent O–H bond.

This causes the equilibrium to lie to the right-hand side. Carboxylic acids are more acidic than alcohols, whose anions are not stabilised by delocalisation.

(b)   For CH_3CH_2OH and $CH_3CH_2NH_2$:

No delocalisation in any of these species.

$$CH_3CH_2OH \rightleftharpoons CH_3CH_2O^- + H^+$$

$$CH_3CH_2NH_2 \rightleftharpoons CH_3CH_2NH^- + H^+$$

For these two, the negative charge is on a different type of atom; it is more stable on the more electronegative atom, O, than on N. Therefore ethanol is the stronger acid.

H & P sections 3.2, 3.5 (pp 33, 36)

Question 6

Draw out the compounds and the equilibria in full. NO_2 is an electron-withdrawing group.

Benzoic acid and 4-nitrobenzoic acid are both carboxylic acids. Assign the pK_a values of 10.2 and 7.2 to the correct acid, giving your reasoning.

....

Answer

Benzoic acid pK_a 10.2; 4-nitrobenzoic acid pK_a 7.2.
$pK_a = -\log_{10}K_a$ so the K_a values are ~10^{-10} and ~10^{-7}; K_a ~10^{-7} and pK_a 7.2 correspond to the stronger acid.

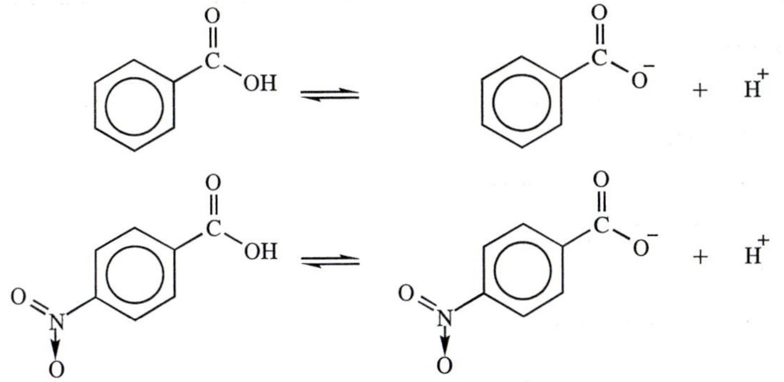

A

The electron-withdrawing NO_2 group will stabilise the negatively charged form **A**, making 4-nitrobenzoic acid the more highly dissociated, stronger acid.

This is an inductive effect of the NO_2 group. Representing the dative bond $N{\rightarrow}O$ as $N^+{-}O^-$ shows the positive charge on nitrogen. This helps to disperse and stabilise a negative charge.

H & P sections 3.2, 3.5 (pp 33, 37)

Question 7

TENORMIN is a drug used in the treatment of high blood pressure, angina and abnormal heart rhythms. Giving your reasons, suggest which type of N-H in this molecule is more acidic.

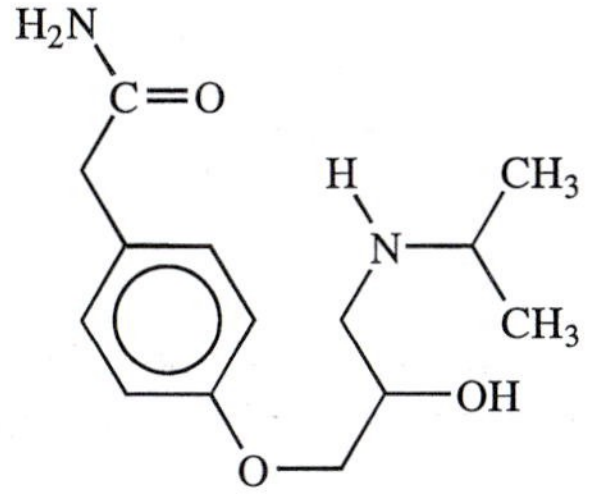

Tenormin

Answer

The two functional groups present which contain N-H are amide and amine.

For an acid $HA \rightleftharpoons H^+ + A^-$

Amides are more acidic than amines because the amide anion, $-CONH^-$, is stabilised by delocalisation (resonance) whereas the amine anion, R_2N^-, is not.

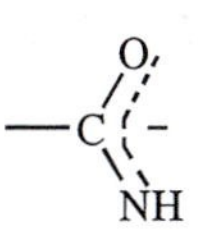

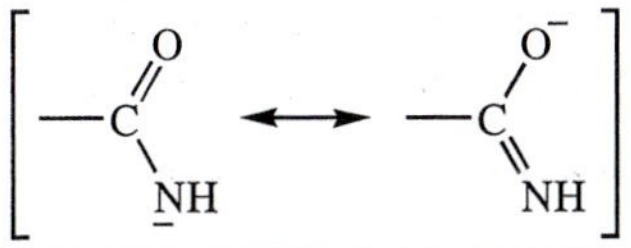

Stabilisation of A^- makes the equilibrium lie more towards H^+ and A^-, which makes HA a stronger acid.

H & P section 3.5 (p 37)

Question 8

PHENTERMINE is a drug used to suppress appetite. Do you think that the pK_b of phentermine is nearer 9 or 4? Use the table and give your reasoning.

Phentermine

Base	pKb
Phenylamine $C_6H_5NH_2$	9.4
Ammonia NH_3	4.8
Trimethylamine $(CH_3)_3N$	4.2
Ethylamine $CH_3CH_2NH_2$	3.2

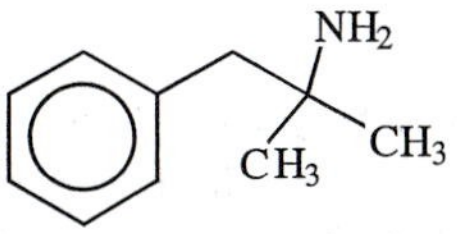

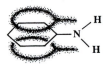

C$_6$H$_5$NHCH$_3$ is classified as an aryl amine but C$_6$H$_5$CH$_2$NH$_2$ as an alkyl amine.

H & P sections 3.2, 3.6 (pp 33, 38)

Answer

Nearer 4. This is because it is an alkyl amine (like trimethylamine and ethylamine) rather than an aryl amine (like phenylamine). The equation for phenylamine is

$$C_6H_5NH_2 + H^+ \rightleftharpoons C_6H_5NH_3^+$$

In C$_6$H$_5$NH$_2$, the nonbonded pair on nitrogen is drawn into the delocalised π system of the benzene ring. This stabilises the molecule by lowering its energy.

This can also be shown with curly arrows: see margin.

If the nonbonded pair is used to make a bond to H$^+$, this energetically favourable stabilisation will be lost. Therefore phenylamine is a weak base with a high pK_b (less protonated).

No such delocalisation occurs in alkyl amines (like ethylamine) which are stronger bases. In phentermine the N is not attached directly to the benzene ring and its nonbonded pair is not delocalised, so it is classified as an alkyl amine.

Question 9

Draw out the structures in full, concentrating on the nitrogen atoms.

Benzylamine, C$_6$H$_5$CH$_2$NH$_2$, and nitrobenzene, C$_6$H$_5$NO$_2$, both contain nitrogen. Explain why benzylamine is a much stronger base than nitrobenzene.

Answer

$$C_6H_5CH_2\overset{..}{N}H_2 + H^+ \rightleftharpoons C_6H_5CH_2\overset{+}{N}H_3$$

The N of benzylamine is not directly attached to the aryl ring; therefore its nonbonded pair of electrons is fully available to bond to H$^+$. C$_6$H$_5$CH$_2$NH$_2$ behaves like an aminoalkane so is quite basic (see answer to Question 8).

The N atom in nitrobenzene does **not** have a nonbonded pair of electrons to bond to H$^+$ because it has used these electrons in a dative bond to oxygen. The structure can be drawn as shown below.

It is a common error to assume that all compounds containing nitrogen atoms are basic.

H & P sections 3.2, 3.6 (pp 33, 38)

Hence the N is non-basic and nitrobenzene is a neutral compound.

Question 10

ATENOLOL is an antidepressant drug. Draw the structure you expect for atenolol in water after one equivalent of HCl has been added, justifying your answer.

Atenolol

'One equivalent' means in the mole ratio 1:1.

Consider the functional groups in turn.

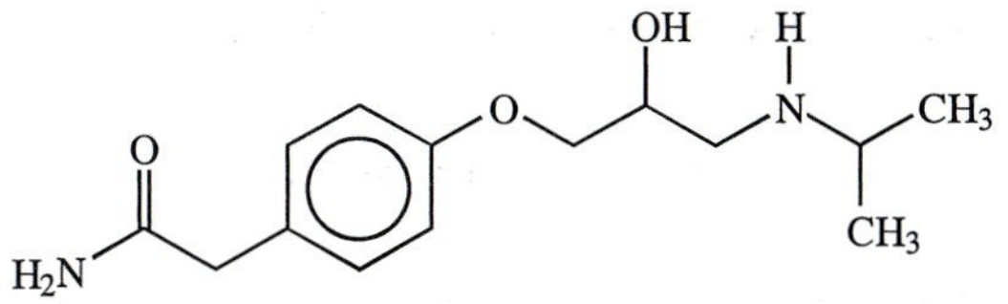

....

Answer

Structure:

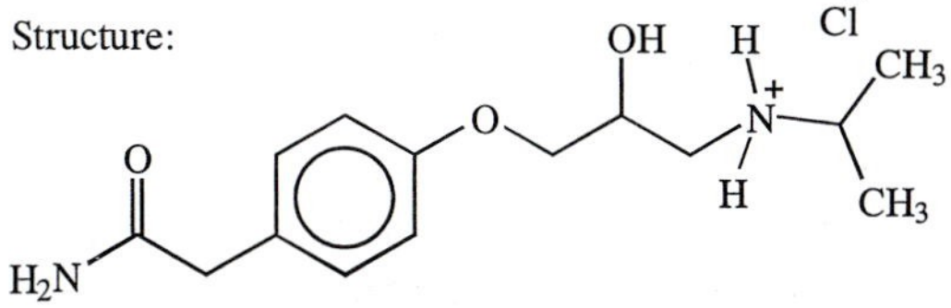

We are adding an acid so we look for the most strongly basic site.

$$B + HCl \rightleftharpoons BH^+ \, Cl^-$$

Pick out the functional groups, and see which is expected to be the best base. We expect amines to be stronger bases than amides or ethers or alcohols. Reasoning:

(a)   The electronegativity of O is greater than that of N, so oxygen atoms exert a stronger pull on nonbonded pairs of electrons than nitrogen. Hence alcohols and ethers are less basic than amines.

(b)   The amide nitrogen nonbonded pair is delocalised with the carbonyl π pair, so it is less available for bonding to a proton than an amine nonbonded pair. Hence amides are less basic than amines.

In resonance terms:

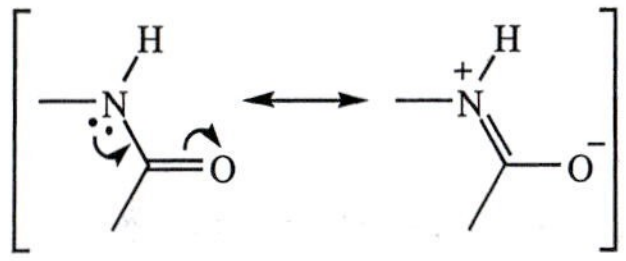

H & P section 3.6 (p 38)

Question 11

The pH indicator Congo Red is a weak acid, $pK_a = 4$. The acid form is blue and the base form is red. Explain why Congo Red is a good indicator for pH changes between 3 and 5.

Look at the answer to Question 3(b). Use calculations in your answer.

....

Answer

For brevity, call Congo Red HIn.

H-In $\rightleftharpoons$ H$^+$ + In$^-$
(blue)   (red)

From the answer to Question 3(b),

$$pH = pK_a + \log \frac{[In^-_{(aq)}]}{[HIn_{(aq)}]}; \quad pK_a = 4$$

At pH 3

$$3 = 4 + \log\frac{[In^-_{(aq)}]}{[HIn_{(aq)}]}; \quad \text{therefore} \quad \log\frac{[In^-_{(aq)}]}{[HIn_{(aq)}]} = -1$$

The 'usable range' of an indicator depends on how well the observer's eye can distinguish the colours.

H & P section 3.2 (p 33)

This ratio In$^-$ (red):HIn (blue) is 1:10 and the solution will look blue.
   Similarly at pH 5, red:blue will be 10:1 and the solution will look red.
   Over the pH change 3 to 5 the indicator changes from red to blue, across its own pK_a of 4.

Question 12

The structures of the amino acids valine, lysine and tyrosine at pH 7 are shown below:

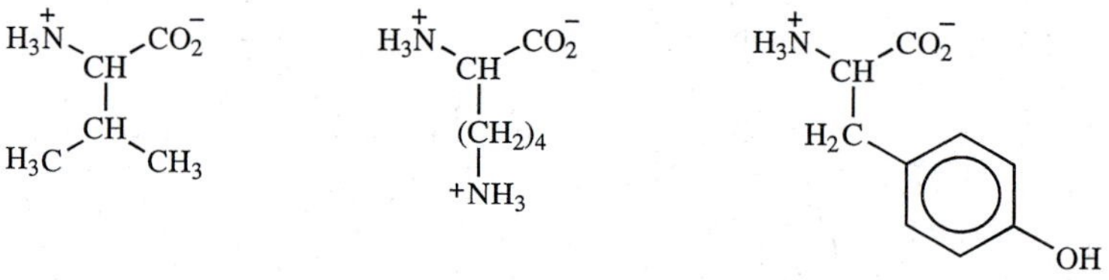

Valine   Lysine   Tyrosine

   Draw diagrams to show the structures you expect for these amino acids at pH 1 and at pH 13.

....

Answer

| pH = 1 (acidic) | pH = 7 (neutral) | pH = 13 (alkaline) |

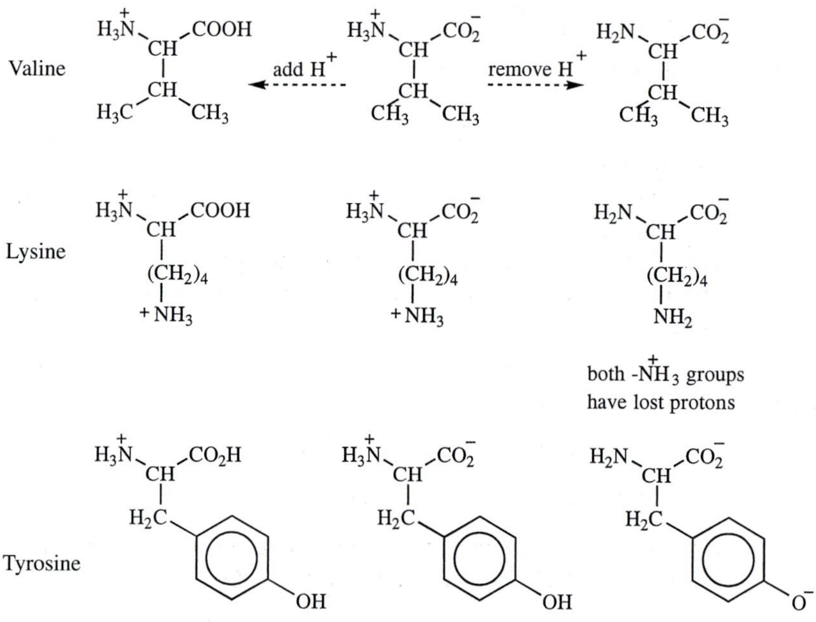

Valine

Lysine

both -$\overset{+}{N}H_3$ groups have lost protons

Tyrosine

the phenolic OH is also acidic

H & P sections 3.4, 3.5, 3.6 (pp 35, 36, 38)

4 Reactions with nucleophiles

4.2 Worked examples

Question 1

(a) Explain why each of the following may be classed as nucleophiles.

   (i) CH_3S^- (ii) $(C_6H_5)_3P$

   (iii) hex-1-ene   (iv)

Look for the ability to donate a pair of electrons for the formation of a new bond. Nonbonded pairs or C=C are usually needed.

(b) Give the equations for
   (i)   one reaction in which OH^- acts as a base towards an organic compound;
   (ii)   one reaction in which OH^- acts as a nucleophile.
   Use your two examples to explain the difference between a base and a nucleophile.

Revise the definitions of base and nucleophile.

....

Answer

(a) (i) $CH_3-\ddot{\underset{\displaystyle ..}{S}}{:}^-$ has three nonbonded pairs.

Elements at the right-hand side of the Periodic Table are most likely to have nonbonded pairs.

   (ii) $(C_6H_5)_3P\!:$ has one nonbonded pair.

   (iii) Hex-1-ene has a π pair in the double bond which can be used to form a new bond, for example to the H of HBr.

(iv) Benzene has three delocalised π pairs, one of which may be donated to form a new bond in the first stage of electrophilic substitution, e.g.

Electrophiles and nucleophiles are complementary: one needs the other.

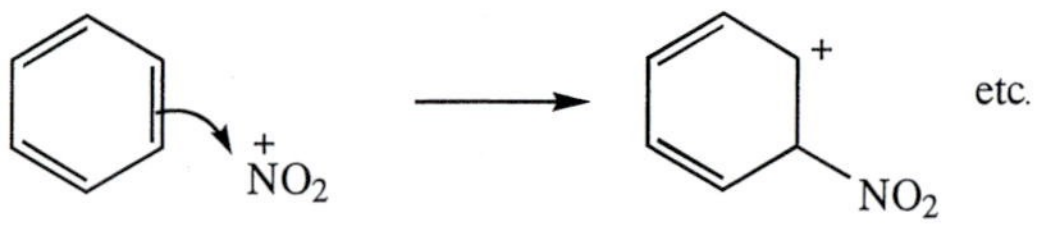

(b) (i) Bases are proton (H^+) acceptors. Any acid/base reaction involving proton transfer will do. Such reactions are equilibria. For example,

$$CH_3COOH + OH^- \rightleftharpoons CH_3COO^- + H_2O$$

(ii) Nucleophiles supply pairs of electrons for the formation of new bonds in reactions. For example,

$$CH_3CH_2CH_2Br + OH^- \longrightarrow CH_3CH_2CH_2OH + Br^-$$

Differences:

1. Basic strength is measured by the *equilibrium constant* of the proton exchange, nucleophilic strength by the *rate constant* of the chosen reaction.

A good nucleophile reacts fast.

2. All bases and most nucleophiles have nonbonded pairs, but for basicity the new bond is always formed to a proton (H^+), whereas for a nucleophile it is to any atom except H, often carbon.

As H is the smallest atom of all, steric effects are usually more important for nucleophiles than for bases.

H & P section 4.1 (p 40)

Question 2

(a) Draw a mechanism and energy profile for the reaction of bromoethane with aqueous KOH. Predict the structure of the transition state for the reaction.

(b) If you were investigating the kinetics of this reaction, how would you:
   (i)   stop the reaction in samples of the reaction mixture;
   (ii) analyse the 'quenched' samples?

What kind of substances are involved as reactants and products? How can any of these be measured quantitatively?

Answer

(a)

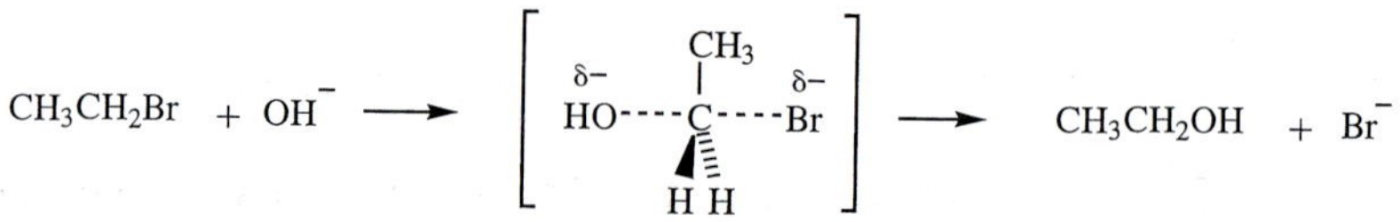

transition state TS

The Hs on the CH_2 are small and do not hinder the approach of OH^-. An alternative S_N1 mechanism would go *via* the very unstable primary carbocation, $CH_3CH_2^+$, and would not be favoured.

Bromoethane is a primary halogenoalkane and its reaction with aqueous alkali is likely to be a single-stage reaction, bimolecular nucleophilic substitution, S_N2.

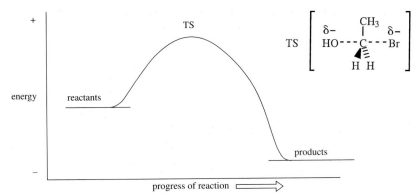

(b) (i) The hydroxide ion can be neutralised with cold dilute nitric acid.

$$\overset{+}{H}_{(aq)} + \overset{-}{OH}_{(aq)} \rightleftharpoons H_2O_{(l)}$$

(ii) The neutralised mixture could be titrated with silver nitrate solution using a potassium chromate indicator. A quantitative precipitate of silver bromide forms, and so the amount of Br^- formed in the sample can be calculated. Alternatively samples of reaction mixture could be quenched with a known *excess* of dilute nitric acid. Back-titration of the remaining nitric acid with standardised alkali would allow calculation of the OH^- concentration left in the original sample.

$$\overset{+}{Ag}_{(aq)} + \overset{-}{Br}_{(aq)} \longrightarrow AgBr_{(s)}$$

H & P section 4.2 (p 41)

Question 3

(a) Draw a mechanism for the reaction of ethanoate ion with iodomethane to give methyl ethanoate and iodide ion.

(b) Compounds **A** and **B** react with ethanoate ion in a similar way. Predict the full structure of the product **C**.

Write the equation first. Look for a nucleophile to displace the I^- ion, and check that the right atoms get connected.

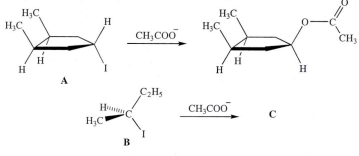

A

B

a single optical isomer

It may help to draw the transition state for this reaction: see Question 2 (and Questions 7 and 8 in Chapter 2).

....

Answer

(a) Equation: $CH_3COO^- + CH_3I \longrightarrow CH_3COOCH_3 + I^-$

This looks like a nucleophile because of its nonbonded pairs and its negative charge.

The CH_3COO^- is drawn in its non-delocalised form because it is easier to follow the electron pair movements with this diagram.

(b) Try a nucleophilic substitution on the iodomethane (S_N2).

The weakness of the C–I bond is very important here.

C is

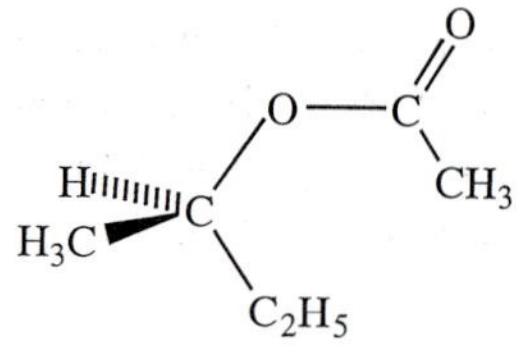

The simplest mechanism for the reaction of **A** is S_N2: ethanoate ion approaches from the top face of **A** as iodide ion leaves from the bottom face. Applying this to **B** and carefully drawing the transition state we can propose a structure for **C**.

This stereochemical change is called an *inversion of configuration* at the asymmetric carbon, and it is characteristic of an S_N2 reaction.

H & P sections 4.1, 4.2 (pp 40, 42)

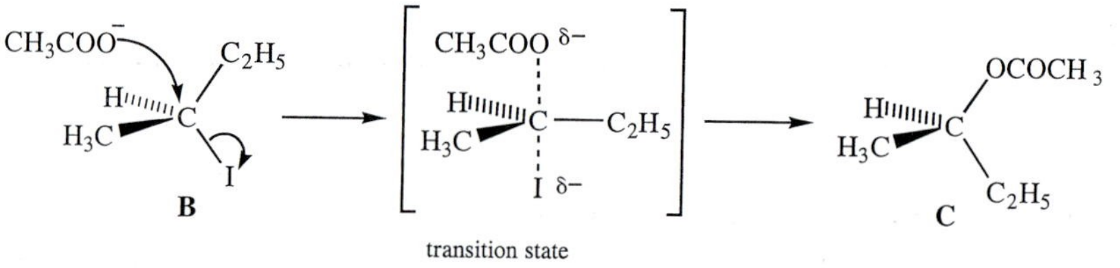

transition state

Question 4

Alkoxide and phenoxide ions react with haloalkanes. Draw a mechanism for the formation of 2,4-D from the phenol **A** and chloroethanoic acid in the presence of NaOH.

What is the function of the sodium hydroxide?

2,4-D is a selective weedkiller for broad-leaved weeds in grassland.

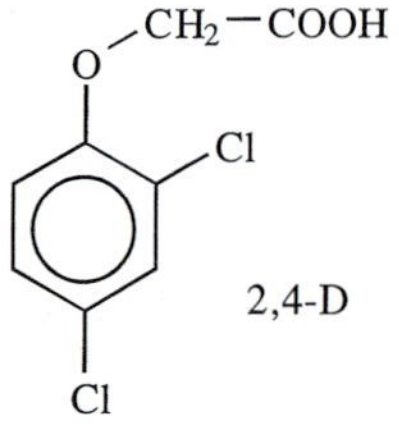

2,4-D

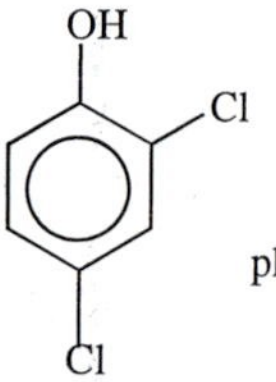

phenol **A**

Answer

Chloroethanoic acid becomes chloroethanoate ion in the presence of alkali.

$$ClCH_2COOH \ + \ OH^- \ \longrightarrow \ ClCH_2COO^- \ + \ H_2O$$

phenate ion of **A**

The phenol will become phenate ion, $ArOH \longrightarrow ArO^-$.

Both anions could act as nucleophiles but only the chloroethanoate possesses a suitable site for nucleophilic attack, the carbon of the C–Cl bond.

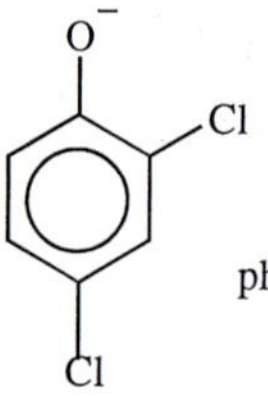

The chlorine atoms on the benzene ring are not susceptible to easy nucleophilic substitution.

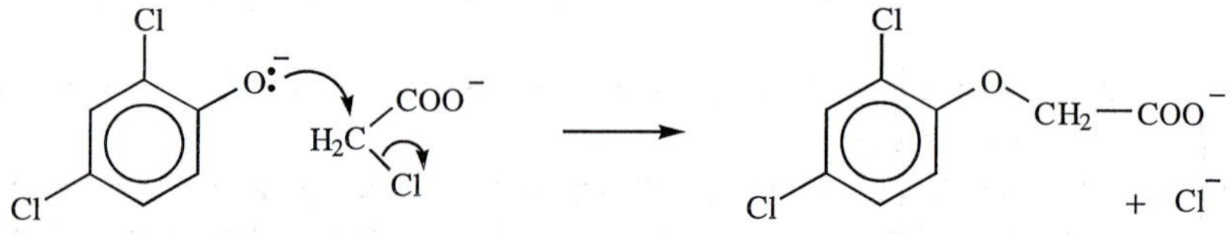

Then acidify to produce the parent acid 2,4-D.

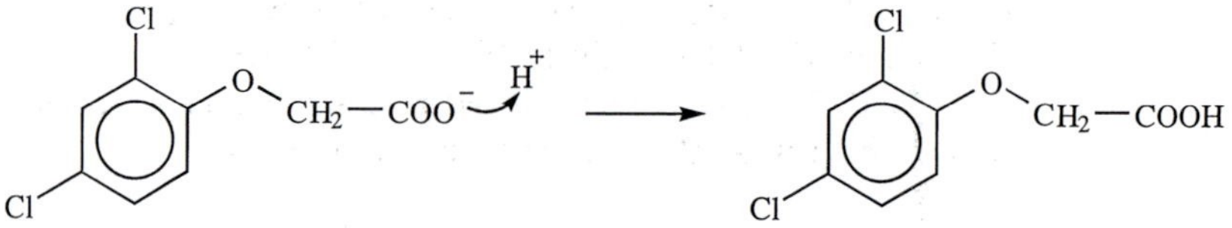

It can help to work backwards by 'disconnecting' the product, so that you can see which atoms in the starting compounds need to be connected to make the product, and which bonds need to be broken. X⌇Y signifies a disconnected bond, for which the double-stemmed arrow ⇒ is used.

This is called the 'disconnection method' or 'retrosynthetic analysis'.

H & P section 4.2 (p 42)

Question 5

(a) Suggest structures for $C_5H_{11}Br$ which do *not* have optical isomers.
(b) Would you expect the products of the reactions of your compounds with cold aqueous sodium hydroxide to show optical isomerism? Explain your answer.
(c) Would you expect the products of the reactions of your compounds with hot alcoholic sodium hydroxide to show geometric isomerism? Explain your answer.

Check the criteria for optical and geometric isomerism.

....

Answer

(a) 1. $CH_3CH_2CH_2CH_2CH_2Br$ ⎫
    2. $(CH_3)_2CHCH_2CH_2Br$ ⎬ primary bromoalkanes
    3. $(CH_3)_3CCH_2Br$ ⎭

    4. $CH_3CH_2CHBrCH_2CH_3$ a secondary bromoalkane

    5. $CH_3CH_2C(CH_3)_2Br$ a tertiary bromoalkane

    In each case there is *no* carbon atom carrying four different groups.

Optical isomers are non-superimposable mirror images of each other. They often possess a carbon atom carrying four different groups, such as in

$$CH_3CH_2CH_2 - \overset{\displaystyle H}{\underset{\displaystyle CH_3}{\overset{|}{\underset{|}{C}}}} - Br$$

(b) No. Substitution of the Br in the bromoalkanes by OH will not introduce chirality where there is none to start with. For example, $CH_3CH_2CH_2CH_2CH_2Br$ gives $CH_3CH_2CH_2CH_2CH_2OH$.

(c) Only pent-2-ene, from 3-bromopentane (see answer (a) 4), will show geometric isomerism.
The conditions encourage elimination, producing alkenes, rather than substitution. The compounds listed in (a) would give the following alkenes:
    1. $CH_3CH_2CH_2CH_2CH_2Br$ gives $CH_3CH_2CH_2CH=CH_2$
    2. $CH_3CH(CH_3)CH_2CH_2Br$ gives $CH_3CH(CH_3)CH=CH_2$
    3. $(CH_3)_3CCH_2Br$ cannot eliminate HBr easily (without rearrangement) as there is no H–C–C–Br sequence of atoms.

The mechanism for this elimination is

It is designated E2: elimination, bimolecular (rate $\propto$ [HO$^-$][RBr]).

H & P sections 4.2, 4.6 (pp 41, 46)

4. $CH_3CH_2CHBrCH_2CH_3$ gives $CH_3CH=CHCH_2CH_3$, which fulfils the criterion that geometic isomerism requires two different groups at each end of a C=C:

cis	*trans*

5. $CH_3CH_2C(CH_3)_2Br$ gives $CH_3CH=C(CH_3)_2$ or $CH_3CH_2C(CH_3)=CH_2$ depending upon which H is lost; the product will be a mixture.

All the alkenes except pent-2-ene have two identical groups at one end or the other of their C=C.

Question 6

The reaction of a halogenoalkane, RBr, with sodium hydroxide in aqueous ethanol gave the following kinetic data.

Run	[RBr] $\times 10^2$ mol dm^{-3}	[OH$^-$] $\times 10^3$ mol dm^{-3}	Relative rate of disappearance of RBr
1	2.4	1.2	1
2	2.4	2.4	1
3	1.2	1.2	0.5
4	3.6	1.2	1.5

Try S$_N$1 and S$_N$2

Overall equation: RBr + OH$^-$ → ROH + Br$^-$

(a) Outline two possible mechanisms for the reaction.

(b) Explain how the data can help you to decide which mechanism is operating.

Answer

(a)

(b) Comparing runs 3 and 1, doubling [RBr] doubles the rate. Therefore the reaction is first order with respect to halogenoalkane.

Comparing runs 1 and 2, doubling [OH$^-$] has no effect on rate. The reaction is zero order with respect to [OH$^-$], and first order overall.

This fits with the S$_N$1 mechanism above, where the rate-limiting (slowest) step involves only halogenoalkane.

Rate = k[RBr]

H & P section 4.5 (p 45)

Question 7

For the reaction

$(CH_3)_3CBr$ + $NaOC_2H_5$ $\xrightarrow[\text{in ethanol}]{}$ $(CH_3)_3COC_2H_5$ + $NaBr$
                                                            major product

the rate is proportional to $[(CH_3)_3CBr]$ but independent of $[NaOC_2H_5]$. The energy profile is drawn below:

This means that the bromoalkane is involved *in or before* the rate-limiting (slowest) step, but the $NaOC_2H_5$ is not.

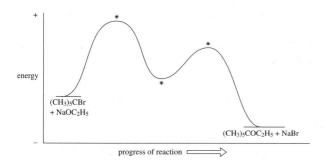

(a) (i) Suggest structures for the organic species at the starred positions.

(ii) Suggest a mechanism for the formation of both major product $(CH_3)_3COC_2H_5$ and minor product $(CH_3)_2C=CH_2$ from the same intermediate.

Think about intermediates and transition states.

(b) Explain the stereochemical results observed in the unimolecular (S_N1) reaction shown below.

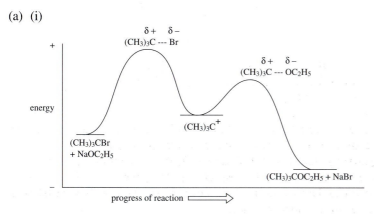

single optical   product ratio 1 : 1
isomer

....

Answer

(a) (i)

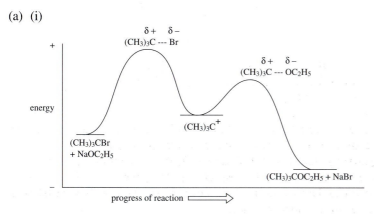

At the central energy minimum there is an intermediate, the carbocation $(CH_3)_3C^+$, with transition states at the energy maxima on each side of it.

Intermediate The reaction rate dependence on $(CH_3)_3CBr$ concentration alone suggests that only $(CH_3)_3CBr$ is involved in the first, slower, rate-limiting step. This could be an ionisation to give the cation $(CH_3)_3C^+$ as the intermediate, which is captured by ethoxide ion in the faster second step.

Transition states (TS) See answer to Question 9 in Chapter 2; here is the method applied to the second TS. The reaction is

$$(CH_3)_3\overset{+}{C} \quad + \quad \,^-OC_2H_5 \quad \longrightarrow \quad (CH_3)_3COC_2H_5$$

The $(CH_3)_3C$ skeleton remains all the way through:

$$(CH_3)_3C$$

The C–O bond is formed in the reaction so in the TS it is partially formed:

$$(CH_3)_3C\text{-}\text{-}\text{-}OC_2H_5$$

The central C's charge changes from $+1$ to 0 so it will be in between, δ^+, in the TS; similarly the oxygen goes from -1 to 0 so will be δ^-. The complete TS is

$$(CH_3)_3\overset{\delta+}{C}\text{-}\text{-}\text{-}\overset{\delta-}{O}C_2H_5$$

and you can work out the structure of the first TS in the same way.

(ii)

The $^-OC_2H_5$ acts as a nucleophile in the formation of the ether but as a base (by removing a proton) in the formation of the alkene. Many factors influence this competition: the type of haloalkane (primary, secondary, etc.), the base/nucleophile used, the solvent, the temperature, etc.

There is competition between overall substitution to give the ether and elimination to give the alkene. In this case both reactions go *via* the carbocation $(CH_3)_3C^+$.

This cation is stabilised by delocalisation (resonance) involving the phenyl ring.

The cation has a plane of symmetry so it is not chiral.

(b) A possible intermediate is $C_6H_5(CH_3)CH^+$. The central positive C has a share in only six outer shell electrons. These three pairs will space themselves as far away from each other as possible: the ion will be trigonal and flat. The chirality of the starting chlorocompound will be lost. Approach of the nucleophile, water, to the cation will be equally easy and equally likely to each face of the flat structure, giving the observed 1:1 ratio of stereoisomers in the product.

$$HO-\underset{\underset{H}{|}\,CH_3}{\overset{C_6H_5}{|}}C\xrightarrow[\text{then loss of H}^+]{\substack{H_2O \text{ adds} \\ \text{to this face}}}\underset{H_3C\ H}{\overset{C_6H_5}{|}}C+\xrightarrow[\text{then loss of H}^+]{\substack{H_2O \text{ adds} \\ \text{to this face}}}\underset{H_3C\ H}{\overset{C_6H_5}{|}}C-OH$$

$$\uparrow\ -Cl^-$$

$$H_3C\underset{H}{\overset{C_6H_5}{\underset{|}{C}}}Cl$$

H & P sections 4.5, 4.6 (pp 45, 46)

Question 8

Draw a transition state for the reaction

$$Nu\bar{:}\;\curvearrowright\;\underset{/}{\overset{\backslash}{C}}\overset{\delta+\ \delta-}{=}O\;\longrightarrow\;Nu-\underset{\backslash}{\overset{/}{C}}\overset{\overset{O^-}{\nearrow}}{}$$

Compare the structures of the starting material and the product. What changes occur?

Answer

$$\underset{/}{\overset{\backslash}{\underset{Nu}{\overset{\delta-}{-}}{---}C}}\overset{\delta-}{=}O$$

This is how to do it.

(1) *Bonds: full bonds*. Put in all the bonds that remain throughout the reaction; here three single bonds, two to C, one to O.

$$\underset{/}{\overset{\backslash}{C}}-O$$

Partial bonds. Bonds that are made or broken during the reaction go in as partial (dotted) bonds. Here the C–Nu is made and the π bond of C=O is broken.

$$Nu----\underset{/}{\overset{\backslash}{C}}==O$$

For example, the TS for CN^- addition to an aldehyde is

$$\overset{\delta-}{NC}----\underset{/}{\overset{\overset{H}{|}\,\backslash}{C}}\overset{\delta-}{==}O$$

and for NH_2NH_2 addition

$$\overset{\delta+}{H_2N-\underset{H_2}{\overset{|}{N}}}----\underset{/}{\overset{\overset{H}{|}\,\backslash}{C}}\overset{\delta-}{==}O$$

(see Question 11)

(2) *Charges*. Put δ^+/δ^- on any atoms which either become charged or lose their charge. Nu goes from minus to no charge, the O from no charge to minus, so both will be δ^-. This completes the transition state.

$$\underset{/}{\overset{\backslash}{\underset{Nu}{\overset{\delta-}{-}}{---}C}}\overset{\delta-}{=}O$$

H & P section 4.7 (p 47)

Question 9

RASPBERRY KETONE (below) is used in synthetic raspberry flavourings. Draw a mechanism for the reduction of this ketone by $NaBH_4$ in ethanol.

This is a nucleophilic addition reaction. Treat $NaBH_4$ as H^- in the mechanism.

Answer

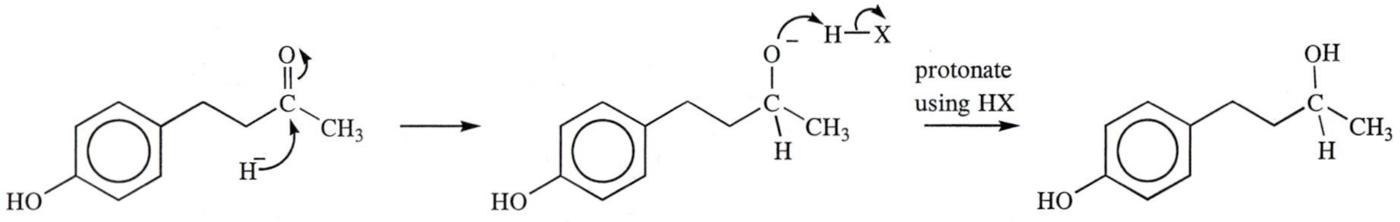

H⁻ represents the
nucleophilic reducing agent.

HX is probably water added
at the end of the reaction

If you can write a mechanism for the BH_4^- reduction of any specific ketone
you can do this one easily by simply copying out your known mechanism in
pencil, such as:

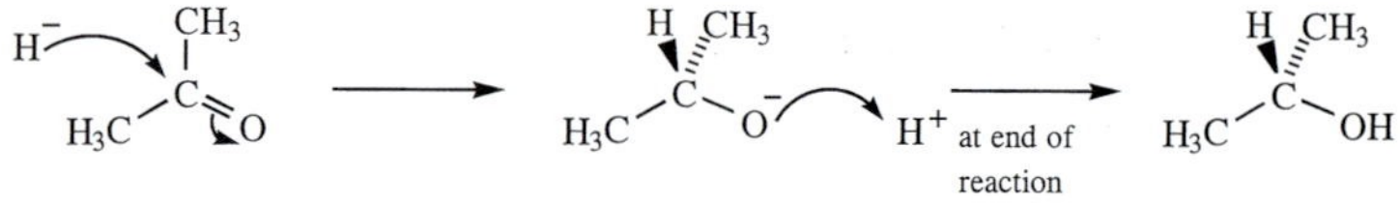

Since the phenolic OH on the ring is
acidic and 'H⁻' is basic, strictly the
mechanism should begin with

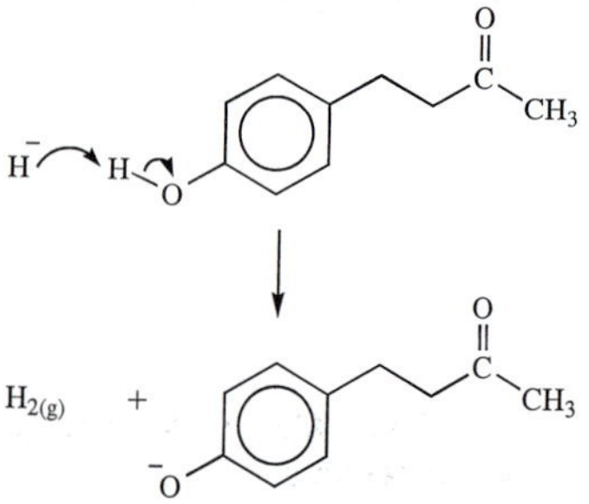

Now rub out *one* of the methyl groups (say the lower one) and replace it by the

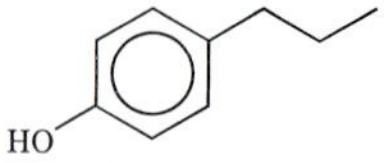

'tail' of raspberry ketone all the way through the mechanism.

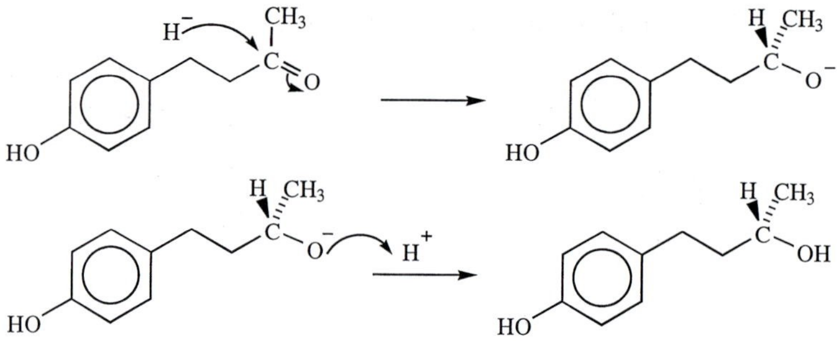

and then continue with this species,
which will be protonated at the end of
the reaction.

H & P section 4.7 (p 47)

Question 10

Write out a mechanism for the reaction below, showing the three stages:
   (i) nucleophilic addition of the N nonbonded pair to the carbonyl group;
   (ii) gain and loss of protons;
   (iii) elimination of water.

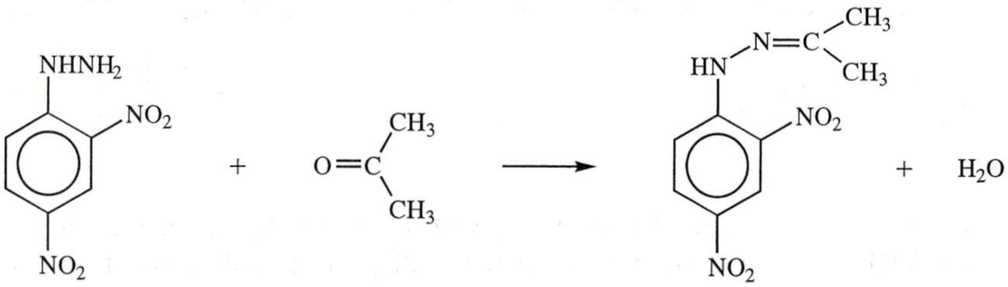

Answer

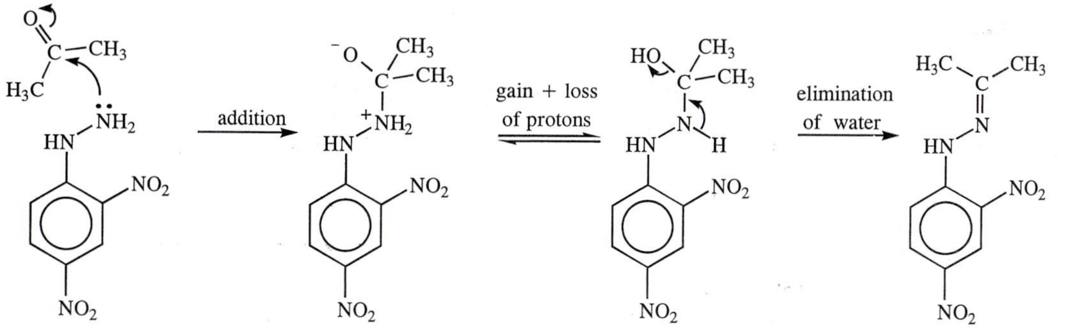

If you know the mechanism for $(CH_3)_2CO + NH_2NH_2$, draw this. Delete one H on NH_2 and replace it by the 2, 4-dinitrophenyl group. The remaining $-NH_2$ uses its nonbonded pair to begin the reaction.

The two N atoms of an aryl hydrazine differ in their nucleophilicity. The nonbonded pair of the N attached directly to the ring is delocalised so this N is less nucleophilic than the terminal N whose nonbonded pair is not delocalised.

H & P section 4.7 (p 49)

Question 11

(a) Predict the structure of the product of the reaction of hydrazine, NH_2NH_2, with lily-of-the-valley aldehyde **A**.

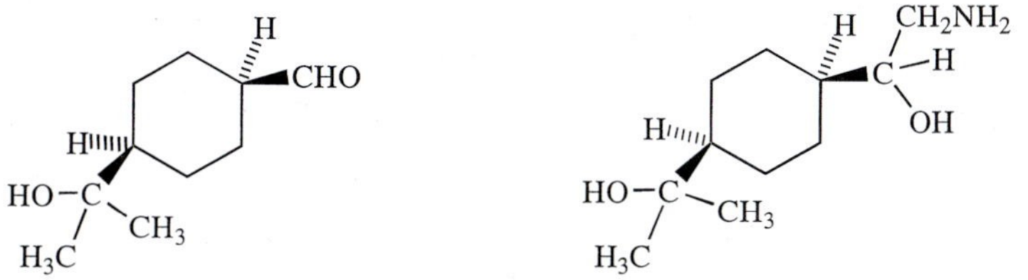

LILY-OF-THE-VALLEY ALDEHYDE is used in perfumery.

(b) Suggest reagents and intermediates for a two-step scheme to convert aldehyde **A** into **B**.

Count the number of C atoms in **A** and **B**: you need one extra.

Answer

(a) Mechanism is:

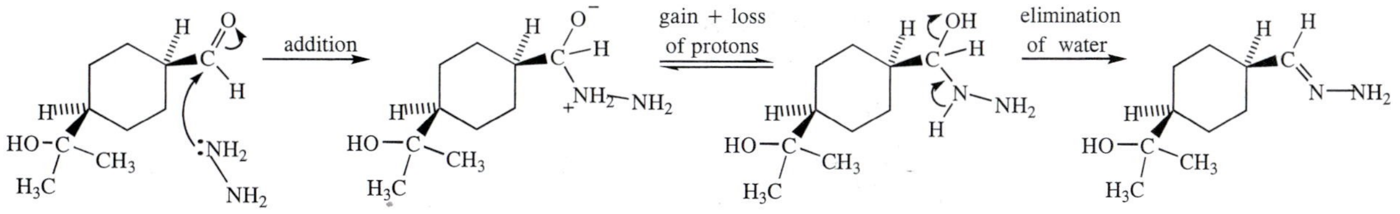

As in Question 12, you can do this by writing out a simpler version, such as $CH_3CHO + N_2H_4$, deleting the CH_3 and writing in the rest of the lily-of-the-valley aldehyde molecule.

(b)

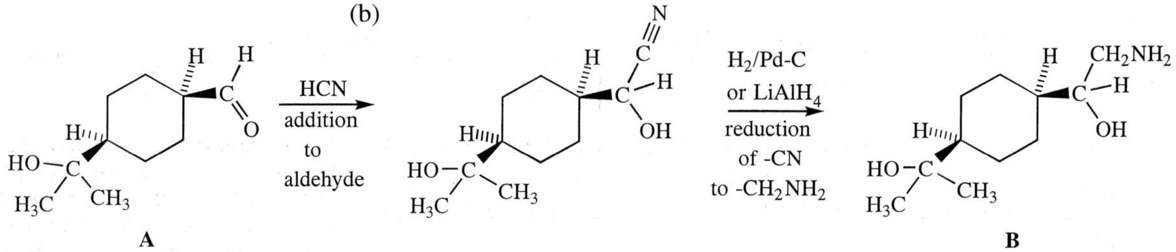

A

H & P section 4.7 (p 49)

Question 12

The reaction of propanone with cold, dilute NaOH gives an unstable intermediate which goes back to propanone.

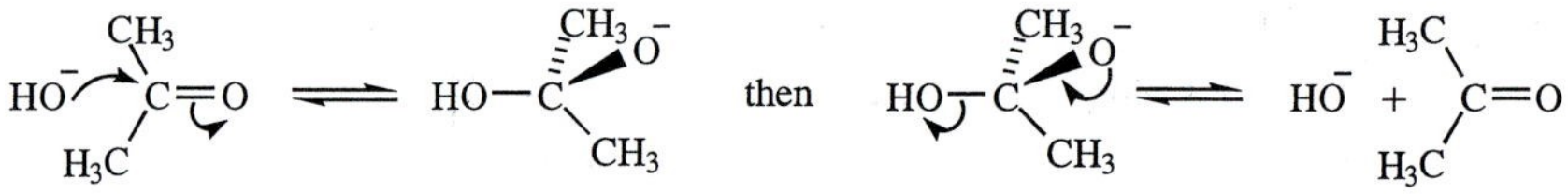

The intermediate can exchange its OH proton with water:

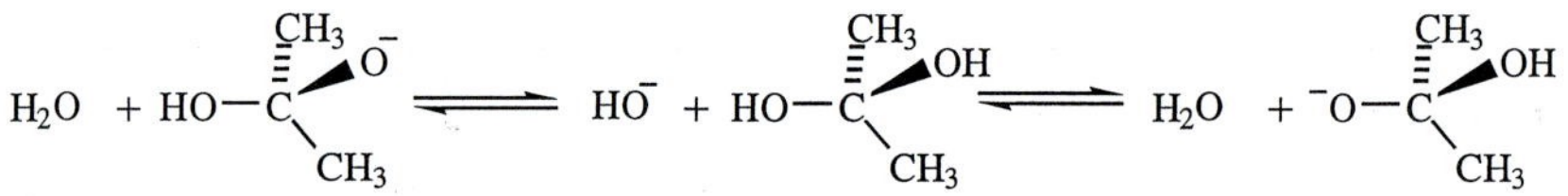

Using this, work out what happens when propanone is treated with cold, dilute NaOH in $H_2{}^{18}O$.

....

Answer

Remember that ^{18}O is simply a heavy isotope of oxygen. It is *not* radioactive.

The propanone becomes labelled with ^{18}O giving $(CH_3)_2C^{18}O$ by the following mechanism. First the ^{18}O is scrambled between the solvent and HO^-:

$$HO^- + H_2^{18}O \rightleftharpoons H_2O + {}^{18}OH^-$$

This $^{18}OH^-$ reacts as above with propanone to give an intermediate, which exchanges its proton and then eliminates $H^{16}O^-$ to give labelled propanone.

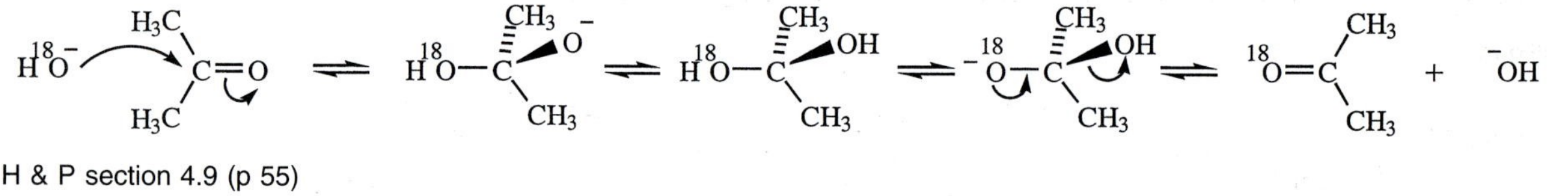

H & P section 4.9 (p 55)

Question 13

Draw the mechanism for the reaction of NaOH with $C_6H_5COOCH_3$.

Answer

This is an ester hydrolysis.

$$C_6H_5COOCH_3 + NaOH \rightarrow C_6H_5COO^-Na^+ + CH_3OH$$

$$C_6H_5-\overset{\overset{\displaystyle O}{\|}}{\underset{\underset{\displaystyle OCH_3}{|}}{C}}\!\!-\!\!{}^-OH \overset{\text{addition}}{\rightleftharpoons} C_6H_5-\overset{\overset{\displaystyle O^-}{\|}}{\underset{\underset{\displaystyle OCH_3}{|}}{C}}\!\!-\!\!OH \overset{\text{elimination}}{\rightleftharpoons} C_6H_5-\overset{\overset{\displaystyle O}{\|}}{C}\!\!-\!\!OH + {}^-OCH_3 \overset{\substack{\text{gain and} \\ \text{loss of proton}}}{\longrightarrow} C_6H_5-\overset{\overset{\displaystyle O}{\|}}{C}\!\!-\!\!O^- + HOCH_3$$

B

Nucleophilic addition to C=O gives the anionic tetrahedral intermediate B, which loses methoxide ion. The strongly basic $^-OCH_3$ ion removes a proton from the carboxylic acid to give the final products, sodium benzoate and methanol.

This process is essentially irreversible because the strong base CH_3O^- removes a proton from the acid to give a carboxylate ion. CH_3OH is too weak a nucleophile to attack the stable anionic carboxylate, so the reaction does not reverse. (Compare with next question.)
H & P section 4.8 (p 52)

Question 14

Complete the mechanism for this acid-catalysed ester hydrolysis:

Overall: $H^+ + CH_3COOCH_2CH_3 + H_2O \rightleftharpoons CH_3COOH + CH_3CH_2OH + H^+$

Answer

The weak nucleophile water adds to the activated protonated carbonyl group. After fast exchange of protons from one oxygen atom to another, elimination of C_2H_5OH together with a proton gives the carboxylic acid. These are all equilibria.

This is a reversible reaction. H^+ can reprotonate the C=O of COOH, then CH_3CH_2OH can add to give a tetrahedral intermediate, and so on.
H & P section 4.8 (p 52)

Question 15

Draw the mechanism for the $LiAlH_4$ reduction of $C_6H_5CH_2OCOCH_3$, which is an active component of oil of jasmine.

This is $C_6H_5CH_2-O-\overset{\overset{\displaystyle O}{\|}}{C}-CH_3$

Answer

Then

H^- is used to represent $LiAlH_4$ in the mechanism.

The products are $C_6H_5CH_2O^-$ and $CH_3CH_2O^-$, so after protonation they will form $C_6H_5CH_2OH + CH_3CH_2OH$.
Overall: $C_6H_5CH_2OCOCH_3 + 4H \rightarrow C_6H_5CH_2OH + CH_3CH_2OH$

H & P sect 4.8 (p 53)

COOH

Benzene-1,4-dicarboxylic acid

Question 16

The polyester TERYLENE can be made by reaction of ethane-1,2-diol and benzene-1,4-dicarboxylic acid.

(a) Draw the structure of the repeating unit of terylene.

(b) What product would you expect if you used ethane-1,2-diamine instead of ethane-1,2-diol?

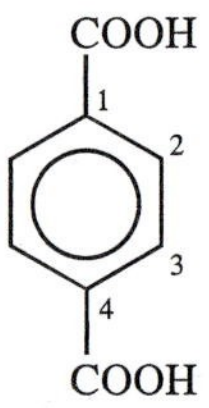

Answer

(a) For example the repeating unit is

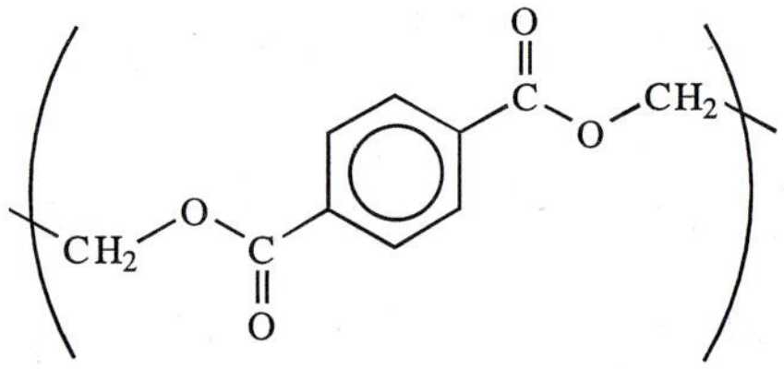

Other ways to draw it include:

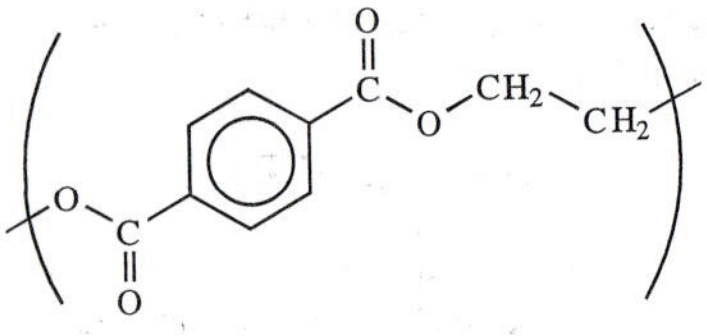

or 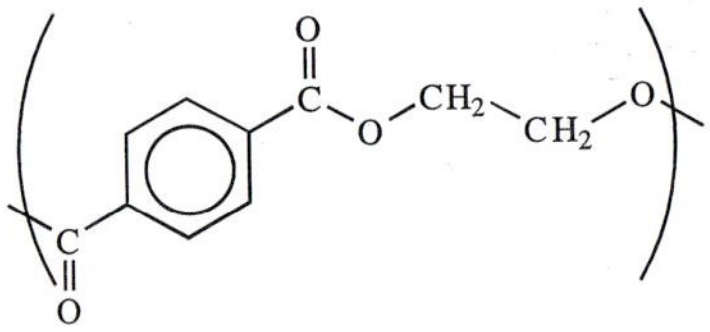   or

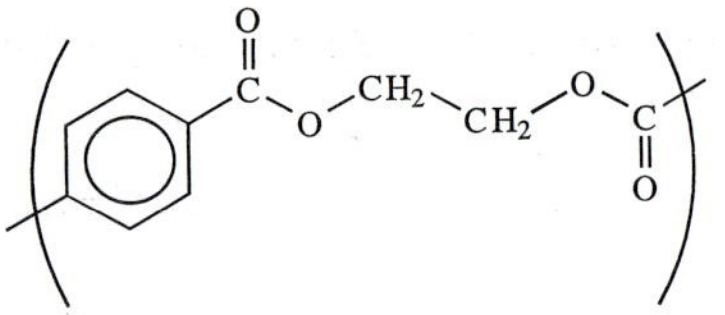

The polyester is made by esterifying both COOH groups with an OH from ethane-1,2-diol. The polymer structure is:

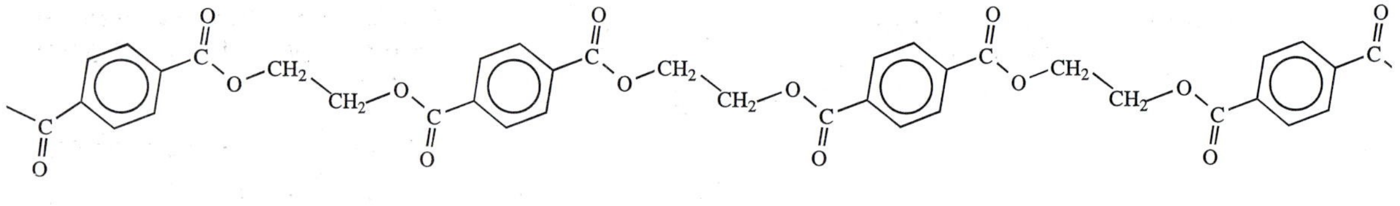

The repeating units laid head-to-tail must make the whole structure.

Polyamides like this have strong chain-to-chain H-bonding. KEVLAR, used in bullet-proof vests, is a similar polyamide.

(b)

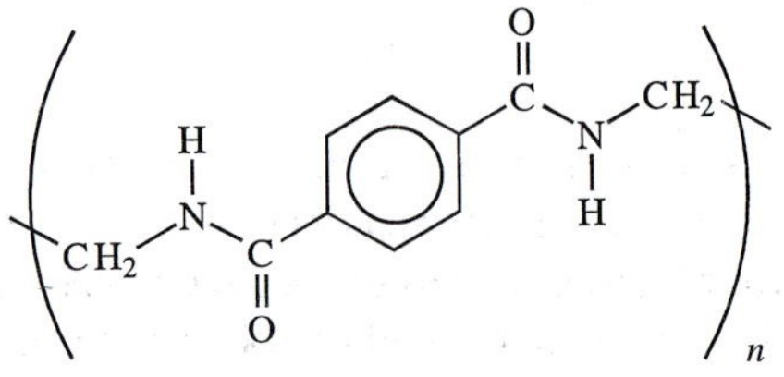

This is a polyamide, made by using the NH_2s as nucleophiles instead of the OHs of the diol.

H & P section 4.8 (p 50)

Question 17

Write reaction schemes showing reagents for these conversions.
(a) $CH_3CH_2OH \rightarrow CH_3CH_2NH_2$
(b) $(CH_3)_2CH\ CH_2CH_2OH \rightarrow (CH_3)_2CH\ CH_2CH_2CN$

More than one step is needed in each case.

....

Answer

(a) It is very difficult to convert the alcohol *directly* into an amine. A stepping stone is needed, and the bromoalkane is a likely candidate.

OH^- is a very poor leaving group in direct substitution reactions.

$$CH_3CH_2OH \xrightarrow{HBr} CH_3CH_2Br$$

$$CH_3CH_2Br \xrightarrow{NH_3 \text{ in ethanol}} CH_3CH_2NH_2$$

HBr is made *in situ* by
$KBr + H_2SO_4 \rightarrow HBr + KHSO_4$

(b) Likewise:

$$(CH_3)_2CHCH_2CH_2OH \xrightarrow{HBr} (CH_3)_2CHCH_2CH_2Br$$

$$(CH_3)_2CHCH_2CH_2Br \xrightarrow{KCN \text{ in aq. ethanol}} (CH_3)_2CHCH_2CH_2CN$$

H & P sections 4.1–4.3 (pp 41–44)

Question 18

Suggest a scheme for the synthesis of compound **B** from cyclohexane carboxylic acid **A**. Several stages will be needed.

Count the carbons in **A** and in **B**, and try using this as an intermediate in your scheme:

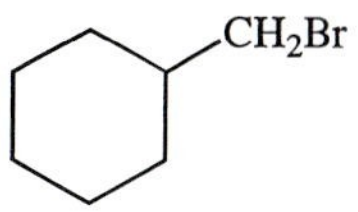

A   B

....

Answer

A possible scheme is

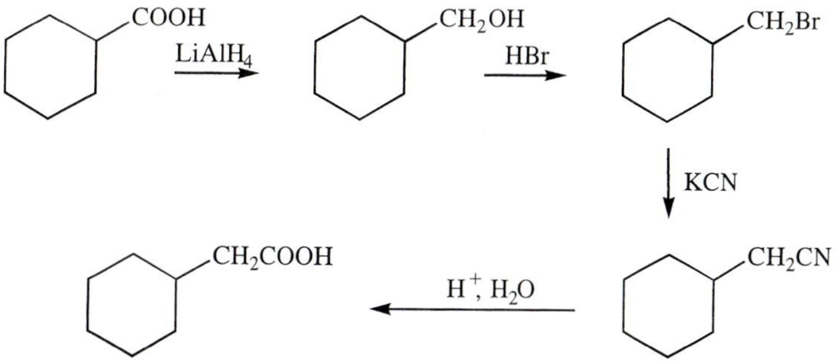

The last two steps can also be achieved neatly by a Grignard reaction:

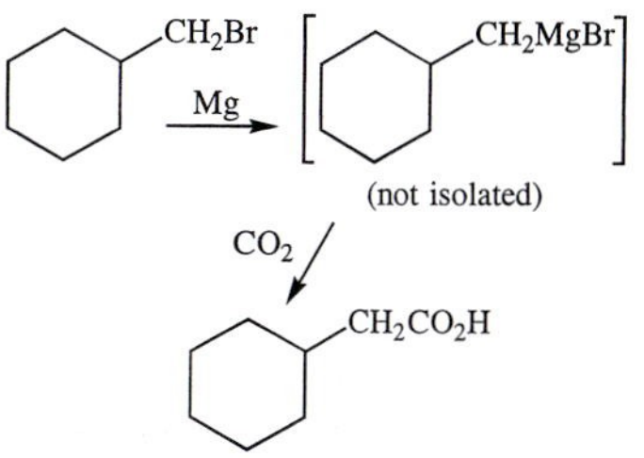

How did we work this out? We will need an extra carbon for the new compound **B**. Using cyanide ion (CN^-) is a way to introduce this. Working backwards, CN^- usually substitutes for a halide, such as the bromide, which is available in turn from the alcohol, a reduction product of the original carboxylic acid **A**.

This is another example of retrosynthetic analysis, denoted by the double-stemmed arrows $\Rightarrow$

5   Reactions with electrophiles

<table>
<tr><td></td></tr>
</table>

Questions by topic:

A: 1, 2, 3, 4, 5, 6, 7

B: 8, 9, 10, 11, 12

C: 13, 14, 15, 16, 17

5.1   What you need to read about or revise for this chapter

A Addition of HX to C=C: including HCl HBr, H_2O addition to ethene, propene, etc. Mechanisms of these reactions. Carbocations and cationic polymerisation.

B Addition of X_2 and XY to C=C: including Br_2, Cl_2, HOBr addition to ethene, propene, etc. Mechanism and stereochemistry of these reactions.

C Electrophilic aromatic substitution reactions: aromatic compounds; halogenation, nitration, acylation. Mechanisms of these reactions.

5.2   Worked examples

Question 1

Assume that the HCl is ionised.

Write out the mechanism for the electrophilic addition of HCl to ethene using both dot diagram and curly arrow versions.

....

Answer

Dot diagram

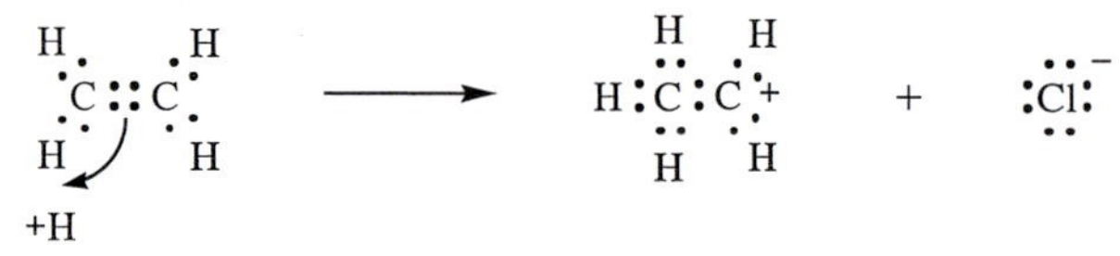

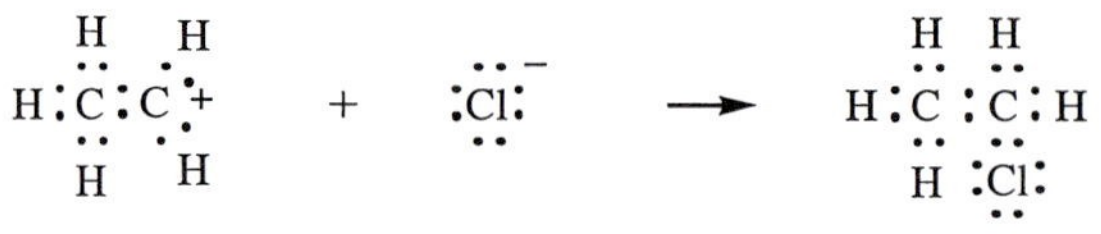

Curly arrow version

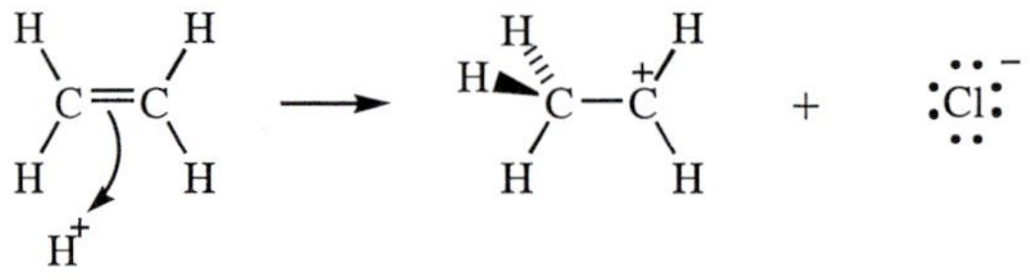

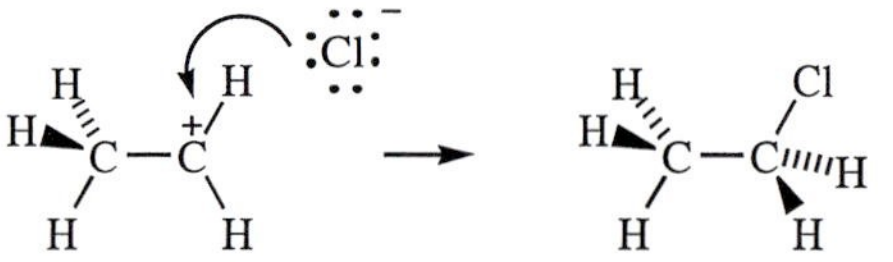

H & P sections 5.1, 5.2 (pp 58,59)

Question 2

Pause at the carbocation stage to assess relative stabilities of the alternatives.

Draw a mechanism for the addition of molecular HBr (not ionised) to propene.

Answer

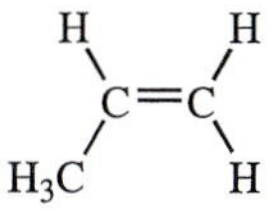

could produce two different carbocation intermediates: either 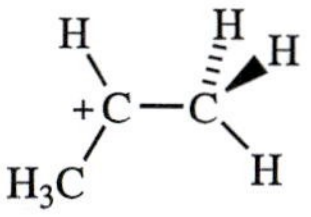   or

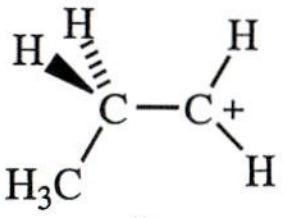

The first of these, the secondary one, will be the more stable of the two.

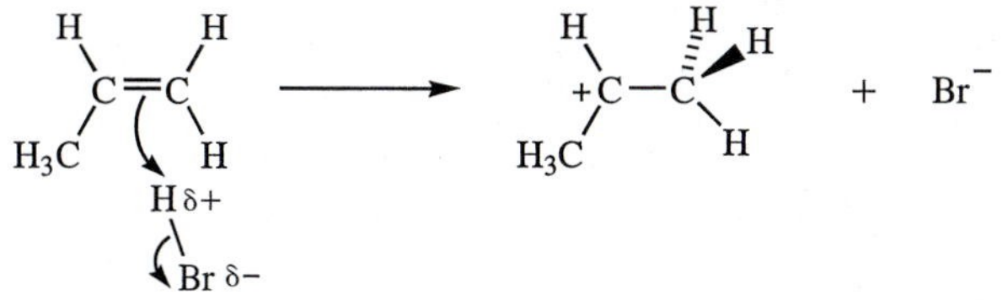

+ Br^-

It is common practice to write the bromide ion as Br^- without all the nonbonded pairs.

then

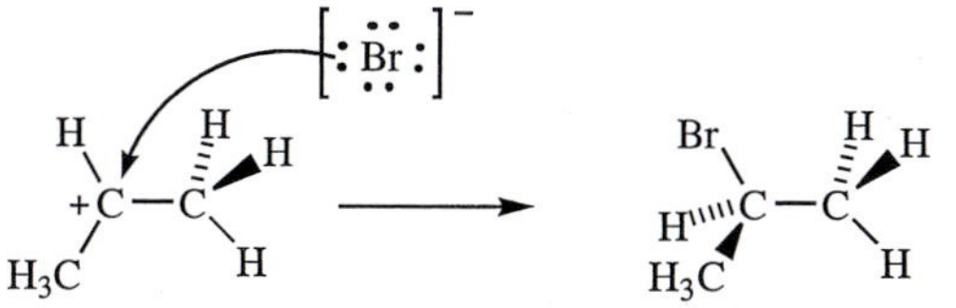

Two effects which contribute to the greater stability of the secondary carbocation are:

(i)   the crowding around the central carbon atom: the larger C–C–C angle (about 120°) in the secondary carbocation keeps the other two carbon atoms further apart than they are in the primary carbocation (C–C–C angle about 109°);

(ii) the electron-releasing inductive effect of *two* alkyl groups on the C^+ favours the secondary C^+, lowering its energy with respect to the primary C^+, which has only *one* carbon substituent.

We normally expect that low-energy intermediates will be preceded by low-energy transition states. Remember that the lower the transition state energy, the faster the reaction will be.

Look up *transition states* and *intermediates* and the relationship between them (H + P, p. 28).

H & P section 5.2 (p 60)

Question 3

Work out the major products of addition of HBr to (a) pent-1-ene and (b) $C_6H_5CH=CH_2$ by considering the relative stabilities of the possible carbocation intermediates in each case.

Answer

(a)

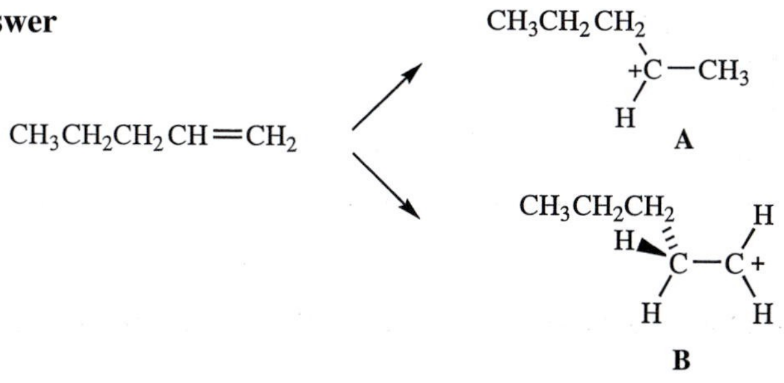

$CH_3CH_2CH_2CH=CH_2$

$CH_3CH_2CH_2$ $+\overset{\cdot}{C}-CH_3$ H   **A**

$CH_3CH_2CH_2$ H $\overset{\cdot}{C}-\overset{+}{C}$ H   H   **B**

Remember that a primary carbocation has *one* carbon substituent on the C^+ carbon, a secondary has *two* and a tertiary carbocation has *three*.

B is a primary carbocation whilst **A** is a secondary one. **A** is the more stable and will give 2-bromopentane, $CH_3CH_2CH_2CHBrCH_3$, as the major product.

(b)

$C_6H_5CH{=}CH_2$

$$
\begin{array}{c}
C_6H_5 \\
\overset{+}{C}{-}CH_3 \\
H \qquad \textbf{C}
\end{array}
$$

$$
\begin{array}{c}
C_6H_5 \quad H \\
H{\cdots}C{-}\overset{+}{C} \\
H \quad\; H \\
\textbf{D}
\end{array}
$$

Cation **C** is more stable and gives $C_6H_5CHBrCH_3$ as the major product. Why?

Cation **C** is secondary, **D** is primary. Much more important is that **C** is stabilised by resonance delocalisation using the aromatic ring.

$$\left[\quad \overset{}{\underset{}{\bigcirc}}{-}\ddot{C}H{-}CH_3 \quad\right]^{+}$$

or, in more detail

$$\left[\; \overset{+}{\bigcirc}\overset{+}{C}H{-}CH_3 \;\longleftrightarrow\; \bigcirc{=}CHCH_3 \;\longleftrightarrow\; \bigcirc{=}CHCH_3 \;\longleftrightarrow\; \bigcirc{=}CHCH_3 \;\right]$$

H & P section 5.2 (p 60)

In **D**, the C^+ is 'insulated' from the π electron system of the aromatic ring by the saturated CH_2 group, and the charge cannot be delocalised.

Question 4

You will need to work backwards from the product to get the starting material (retrosynthesis), and then to write the synthesis out forwards.

Addition of HBr to alkenes gives bromoalkenes. Suggest structures for the alkenes which by addition of HBr would give the following bromocompounds. Draw the intermediates and give your reasoning. (There may be more than one possible alkene for each).

(a) bromocyclohexane   (b) 2-bromo-2-methyl propane
(c) 2-bromobutane   (d) 1-bromo-1-methyl cyclopentane

....

Answer

(a)

(b)

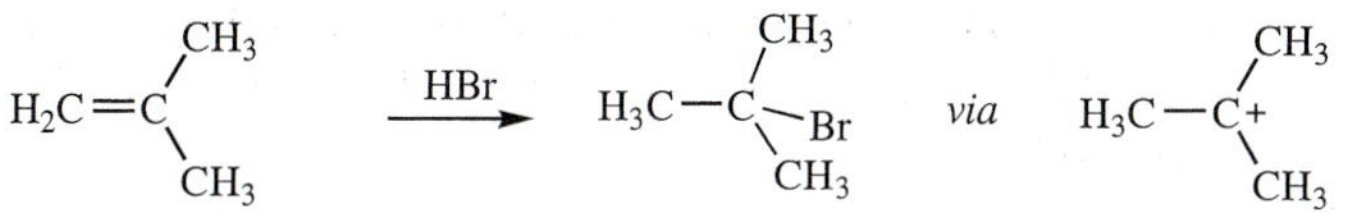

(c)

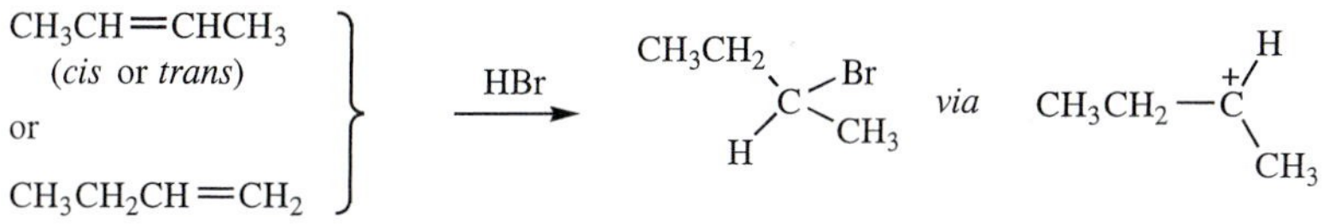

(d)

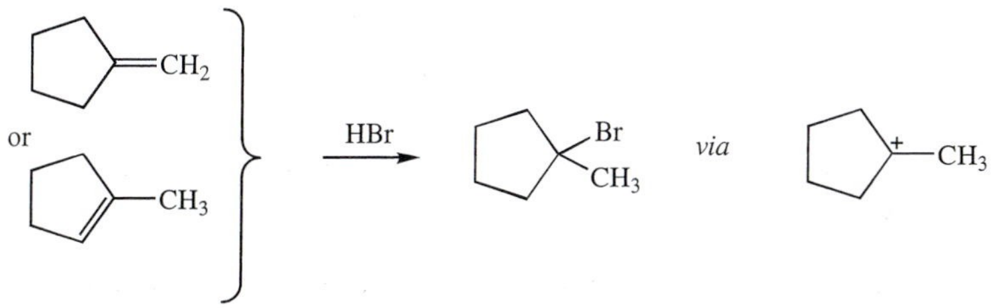

Reasoning   In each mechanism the C=C is first protonated by H$^+$ to give a carbocation; the Br$^-$ then adds to the positive C. So the intermediate has a C$^+$ where product will have the Br attached; e.g. for (d).

Remember that retrosynthetic reasoning is shown with $\Rightarrow$ arrows

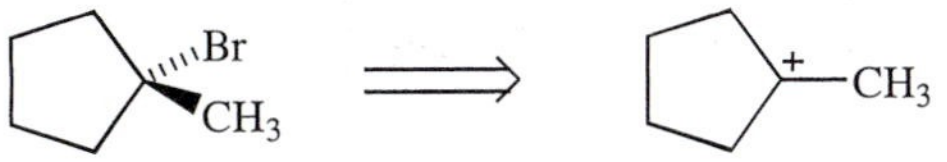

Working backwards, the C$^+$ is formed by addition of H$^+$ to the other end of a C=C: there are two possibilities here.

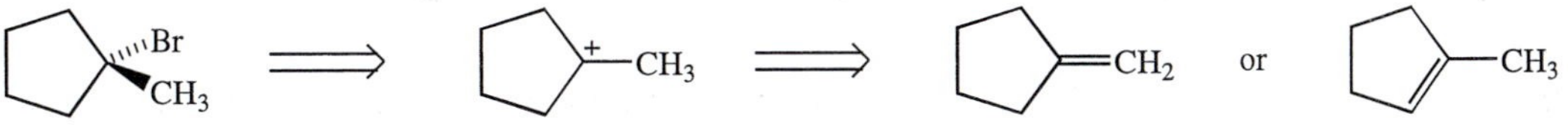

Check your answers by writing the mechanism forwards.

H & P section 5.2 (p 60)

Question 5

Butan-2-ol contains an asymmetric carbon atom, but the product of the reaction of but-1-ene with concentrated H_2SO_4 followed by water is not optically active. Explain.

What are the shapes of but-1-ene and butan-2-ol?

....

Answer

Butan-2-ol has two optical isomers and the optical isomers must be made in equal amounts if the sample shows no optical activity.

But-1-ene is flat around the double bond so that the initial attack by the electrophile H$^+$ can take place from either side, producing the same flat carbocation each time.

$$H_2SO_4 \rightleftharpoons H^+ + {}^-OSO_3H$$

A general rule is that if there is no chirality in any of the starting materials the product mixture will not rotate the plane of plane polarised light.

CH₃CH₂ $\cdots$C=C$\cdots$H with H, H and H⁺

or

H⁺ with CH₃CH₂ $\cdots$C=C$\cdots$H

} ⟶ $CH_3CH_2\cdots\overset{+}{C}-CH_3$ with H

The **flat** carbocation can be attacked by the nucleophile HSO_4^- equally from either side to give the two non-superimposable mirror image isomers.

A 1:1 mixture of optical isomers is known as a racemic mixture.

$^-OSO_3H$ → $CH_3CH_2\cdots\overset{+}{\underset{H}{C}}-CH_3$

⟶ $CH_3CH_2\overset{OSO_3H}{\underset{H}{C}}CH_3$ $\xrightarrow{H_2O}$ $CH_3CH_2\overset{OH}{\underset{H}{C}}CH_3$

or

$CH_3CH_2\cdots\overset{+}{\underset{H}{C}}-CH_3$ $^-OSO_3H$

⟶ $CH_3CH_2\overset{H}{\underset{OSO_3H}{C}}CH_3$ $\xrightarrow{H_2O}$ $CH_3CH_2\overset{H}{\underset{OH}{C}}CH_3$

H & P section 5.3 (p 62)

Question 6

Draw the mechanism and the structure of the product of the cationic polymerisation of $H_2C=C(CH_3)_2$.

....

Answer

Mechanism

$CH_2=\overset{CH_3}{\underset{}{C}}-CH_3$ with H^+

⟶

$CH_3-\overset{CH_3}{\underset{}{\overset{+}{C}}}-CH_3$ $CH_2=\overset{CH_3}{\underset{}{C}}-CH_3$

↓

$H_3C-\overset{CH_3}{\underset{CH_3}{C}}-CH_2-\overset{CH_3}{\underset{}{\overset{+}{C}}}-CH_3$ $CH_2=\overset{CH_3}{\underset{}{C}}-CH_3$

↓

etc. ⟵ $H_3C-\overset{CH_3}{\underset{CH_3}{C}}-CH_2-\overset{CH_3}{\underset{CH_3}{C}}-CH_2-\overset{CH_3}{\underset{}{\overset{+}{C}}}-CH_3$

Pressure-sensitive adhesives can be made from this polymer. The polymer made by co-polymerising $H_2C=C(CH_3)_2$ with $H_2C=CHC(CH_3)=CH_2$ is 'butyl rubber' and is used in making inner tubes of tyres.

Both carbocation intermediates shown are tertiary: more stable than the alternative primary carbocations.

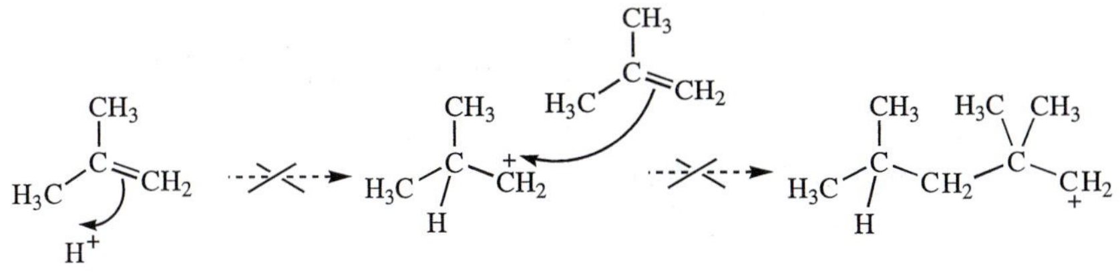

H & P section 5.5 (p 64)

The *structure of the product* is

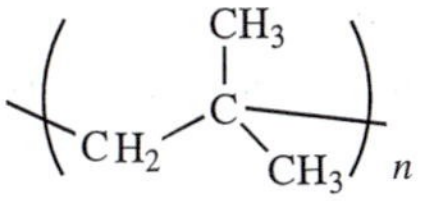

$$\left(CH_2-\underset{\underset{CH_3}{|}}{\overset{\overset{CH_3}{|}}{C}} \right)_n$$

Question 7

(a) Illustrate the meaning of the terms *hydration* and *hydrogenation* using propene as your example.

(b) Suggest a mechanism for the following reaction; $HClO_4$ is a strong acid.

$$CH_3CHCH_2 \xrightarrow{\text{aq. } HClO_4} CH_3CHOHCH_3$$

....

Answer

(a) *Hydration* means the addition of water (H_2O) to a compound e.g.

$$CH_3CH{=}CH_2 \;+\; H_2O \xrightarrow{\;H^+\;} CH_3CHOHCH_3$$

$$C_3H_6 \xrightarrow{+H_2O} C_3H_8O$$

Hydrogenation means the addition of hydrogen (H_2) to a compound e.g.

$$CH_3CH{=}CH_2 \;+\; H_2 \xrightarrow[\text{catalyst}]{\text{Pd/C or Ni}} CH_3CH_2CH_3$$

$$C_3H_6 \xrightarrow{+2H} C_3H_8$$

(b)

$$HClO_4 \;\rightleftharpoons\; H^+ \quad ClO_4^-$$

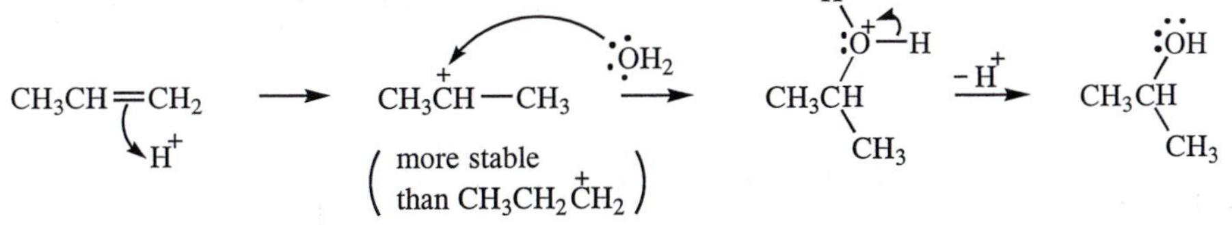

$$CH_3CH{=}CH_2 \xleftarrow{H^+} \quad CH_3\overset{+}{C}H-CH_3 \quad \left(\begin{array}{c} \text{more stable} \\ \text{than } CH_3CH_2\overset{+}{C}H_2 \end{array} \right) \quad \rightarrow \quad CH_3\overset{+}{C}H\underset{CH_3}{\overset{H}{\overset{|}{\underset{|}{O}}}}H \xrightarrow{-H^+} CH_3\underset{CH_3}{\overset{|}{C}H}\ddot{O}H$$

As a strong acid, $HClO_4$ is almost completely dissociated. The C=C is protonated and the intermediate reacts with the abundant nucleophilic water. Loss of a proton (probably to the solvent, water) ends the reaction.

An aqueous solution of any strong oxyacid can be used to hydrate an alkene.

Question 8

Addition of bromine to alkenes often takes place *via* a cyclic bromonium ion rather than a carbocation.

(a)   Draw dot diagrams for the two possible intermediates in the addition of bromine to ethene.

(b)   Do you think that chlorine is more or less likely to form such a cyclic 'onium ion during halogen addition to alkenes? Explain your answer.

Answer

(a)

Carbocation

this C has only 6 electrons in its outer shell

The complete outer shells is a reason why this bromonium ion is more stable than the carbocation above.

Bromonium ion

here both C atoms have 8 electrons in their outer shells

(b)   *Less* likely. The electronegativity of Cl is greater than that of Br, so Cl is *less* likely to share a nonbonded pair.

H & P section 5.4 (p 62)

Question 9

(a)   Write a mechanism for the reaction below:

(b)   Explain why the major product in (a) is the bromo-alcohol and not the

$$CH_2{=}CH_2 \xrightarrow{\quad Br_2 + H_2O \quad} CH_2OHCH_2Br$$

dibromide.

(c)   What additional product would you expect to find if the reaction mixture in (a) also included sodium ethanoate?

Answer

(a)

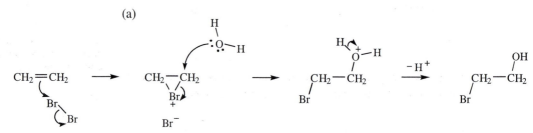

A mixture of Br_2 and water is acidic, so H_2O is the nucleophile and not OH^-.

(b)   In (a) there are two possible nucleophiles to react with the bromonium ion, H_2O and Br^-. Both are moderate nucleophiles but water (the solvent) is present in vast excess, so is statistically more likely to react. Hence the major product is the bromo-alcohol.

(c) The additional product will be $CH_2BrCH_2OCOCH_3$, formed by reaction of ethanoate ion with the cationic intermediate.

H & P section 5.4 (p 63)

Question 10

(a) Write a mechanism for the reaction of propene with bromine, explaining any isomerism that arises.

(b) Account for the formation of $CH_2BrCH(OCH_3)CH_3$ when propene reacts with bromine in methanol.

Which cationic intermediate is formed?

Consider how the methanol gets involved and why the OCH_3 goes onto the second carbon atom.

Answer

(a)

Let the bromine attack from below. This cationic intermediate is not symmetrical and carries more positive charge on the secondary carbon than on the primary.

The $\delta+$ carbon is now attacked by nucleophilic Br^-.

The middle C in the molecule is chiral. The other optical isomer would have been formed if the initial attack by bromine had been from above, and by the nucleophile from below. Both pathways are equally likely, and we get a 1:1 mixture.

(b) Initial electrophilic attack by Br_2 leads to an unsymmetrical cationic intermediate, as before.

This can react with any handy nucleophile, in this case Br^- or CH_3OH. The nucleophile attacks the more positive secondary carbon.

Work this through for yourself. If you cannot see that the two dibromides

are non-superimposable mirror images, make some models to convince yourself.

In the methanolic solution there will be very much more CH_3OH than Br^-.

H & P section 5.4 (p 63)

Question 11

The C_5 rings of cyclopentene and cyclopentane can be taken as flat.

(a)   Carefully draw in 3D the mechanism for the addition of Br_2 to cyclopentene. Show that the product consists of only one *geometric* isomer.

(b)   Now show why the product still consists of two *stereo*isomers.

....

Answer

(a)

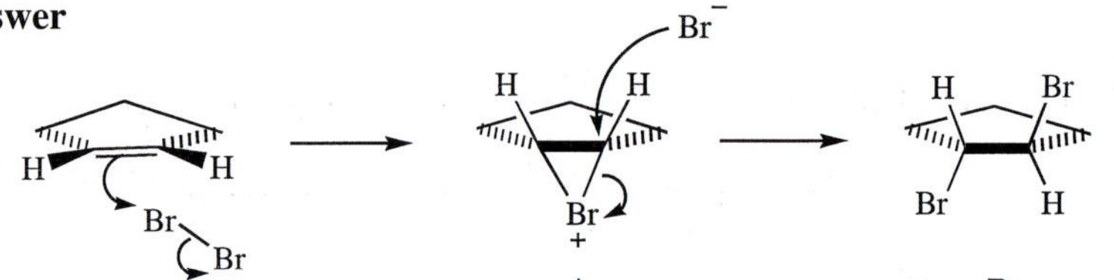

A   B

The two bromine atoms in **B** are on opposite faces of the cyclopentane ring. The other geometric isomer **C** (*not* formed) has the two Br atoms on the same face:

This can also be drawn with initial attack of Br_2 on the top face—this would give **B′** in place of **B** (see answer to (b) below). Check these using models.

B   C

Interconversion of **B** and **C** is prevented by the ring's rigidity.

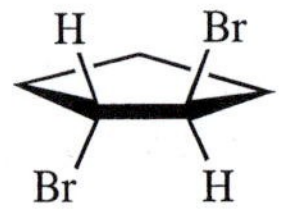

B and **C** are structural isomers that cannot be interconverted because of the restricted rotation about the ring C – C bond: so they are *geometric* isomers. **B** is the *trans* and **C** is the *cis* isomer.

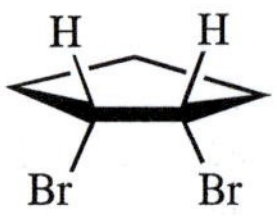

(b)   The other category of stereoisomers is optical. **B** is non-superimposable upon its mirror image so it has two optical isomers, **B** and **B′**. **B′** is formed by reaction of Br^- at the other carbon atom of intermediate **A**.

The attack of Br^- is equally likely at each carbon, so equal amounts of **B** and **B′** will be produced.

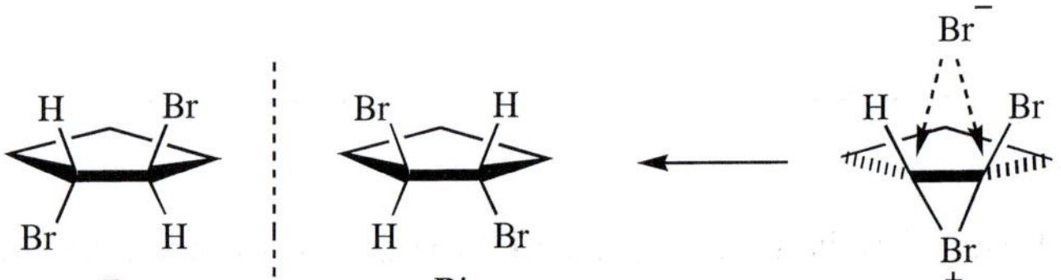

B   B'

Neither **B** nor **B′** has a plane or centre of symmetry. They will rotate the plane of plane polarised light equally but in opposite directions, so that in a 1 : 1 (racemic) mixture the overall rotation will be zero.

H & P section 5.4 (p 62)

Question 12

Suggest reagents for conversion of pent-1-ene into:

Draw the structures of the starting material and the product. Compare them and note the differences and similarities.

(a)   pentan-2-ol;
(b)   1,2-dibromopentane;
(c)   1-bromo-2-hydroxypentane.

Answer

Pent-1-ene is $CH_3CH_2CH_2{}^2CH={}^1CH_2$

(a) $CH_3CH_2CH_2\overset{2}{CH}=\overset{1}{CH_2}$ $\xrightarrow[\text{+ H}_2\text{O}]{\text{aq. H}_2\text{SO}_4}$ $CH_3CH_2CH_2\overset{2}{\underset{\underset{1}{CH_3}}{CH}}\diagup\text{OH}$

Protonation of pent-1-ene gives mostly the more stable secondary carbocation rather than the less stable primary carbocation.

(b) $CH_3CH_2CH_2\overset{2}{CH}=\overset{1}{CH_2}$ $\xrightarrow[\text{+ 2Br}]{\text{Br}_2}$ $CH_3CH_2CH_2\overset{2}{\underset{\underset{1}{CH_2Br}}{CH}}\diagup\text{Br}$

(c) $CH_3CH_2CH_2\overset{2}{CH}=\overset{1}{CH_2}$ $\xrightarrow[\text{+ HOBr}]{\text{aq. Br}_2}$ $CH_3CH_2CH_2\overset{2}{\underset{\underset{1}{CH_2Br}}{CH}}\diagup\text{OH}$

The reagent is bromine water. Bromine is the electrophile and then water adds to the cationic intermediate.

H & P section 5.2, 5.3, 5.4 (p 60, 62)

Question 13

How many different compounds are drawn below? Explain your answer, giving the systematic names.

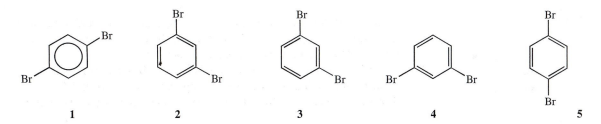

| 1 | 2 | 3 | 4 | 5 |

Answer

Two: 1,3-dibromobenzene (**2**, **3** and **4**) and 1,4-dibromobenzene (**1** and **5**). The 'nondelocalised' and 'hexagon + circle' representations of the aromatic ring are equivalent. Because of the delocalisation, it does not signify where the double bonds are drawn in the nondelocalised forms, so **2** and **3** represent the same compound.

The non-delocalised representations

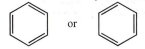

are called the *Kekulé* forms after the man who first proposed them. They are used for mechanisms involving the aromatic ring system of π electrons.

H & P section 5.7 (p 67)

Question 14

Give chemical evidence to support the following statements.
(a) The alkene double bond of phenylethene is more easily oxidised than the aromatic ring.
(b) Phenol is more reactive towards electrophiles than benzene.

Answer

(a)   Alkenes are readily oxidised by $KMnO_4$ but the aryl ring is not affected.

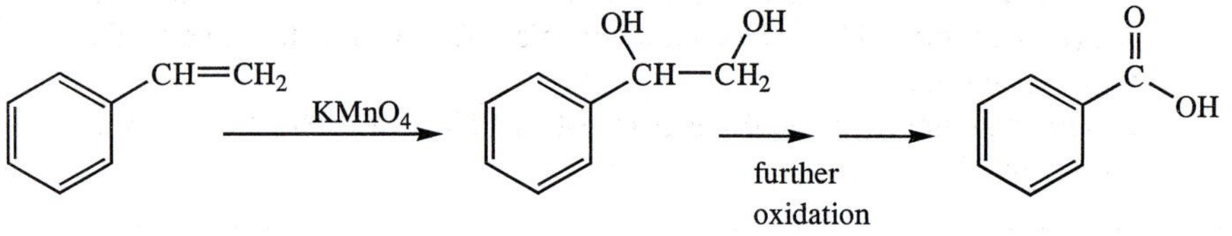

OR

'One molar equivalent' means using the same number of moles of ozone and of phenylethene.
Instead of zinc, other reducing agents can be used, e.g. $(CH_3)_2S$ or hydrogen with a metal catalyst.

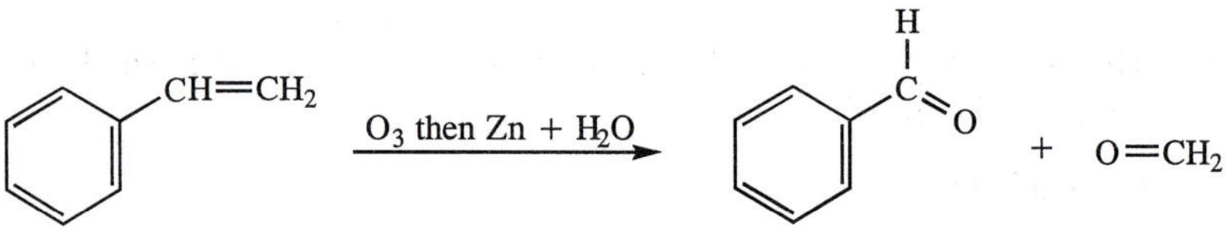

It is also possible to ozonolyse just the alkene $C=C$ if you use only one molar equivalent of ozone.

(b)   Reaction with bromine: benzene needs a catalyst ($FeBr_3$) and heat and gives monobromobenzene; phenol reacts with bromine and water without a catalyst to give tribromophenol.

$FeBr_3$ can be made from iron filings and bromine. $AlBr_3$ is also a good catalyst for this reaction.

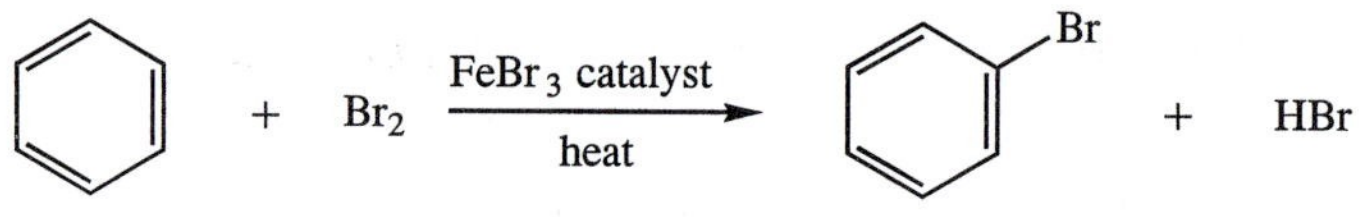

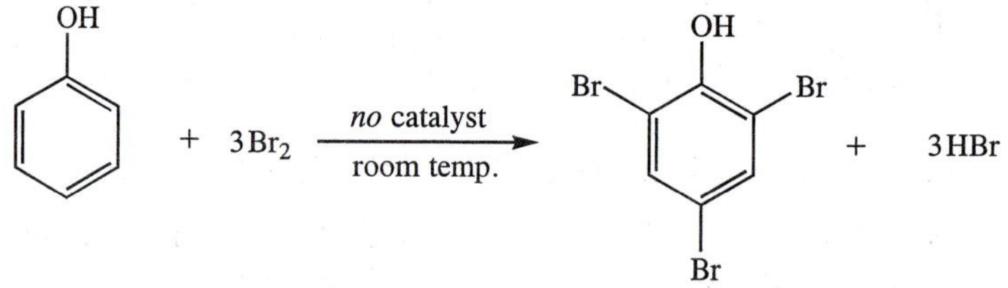

OR

Benzene diazonium ion ($C_6H_5N_2^+$) is a weak electrophile and will react with phenol but not with benzene. It is best if the phenol is in alkaline solution.

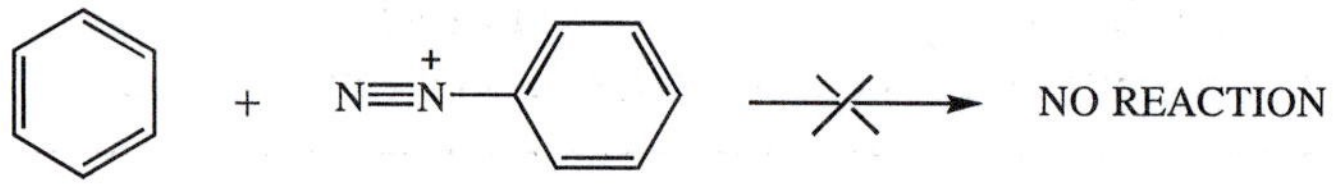

This product is an azo compound as it has the $-N=N-$ grouping. These compounds are important as dyes and as pharmaceuticals.

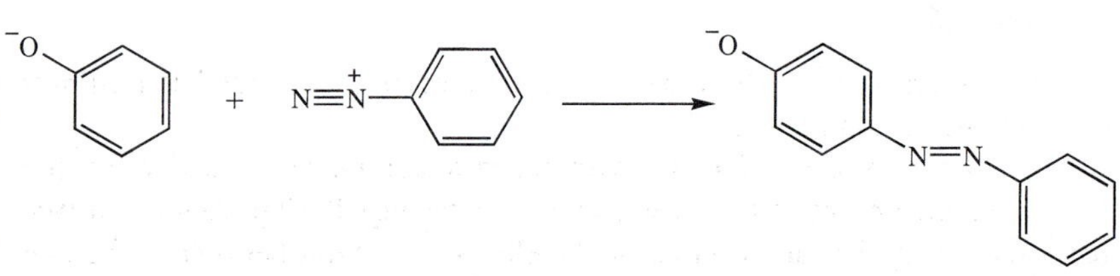

H & P section 5.6, 5.7, 5.8, 5.9 (pp 66, 71, 74)

Question 15

(a) $AlBr_3$ can be used as a catalyst for the bromination of benzene. Why is $AlBr_3$ described as 'electron-deficient'?

(b) In the reaction of benzene, bromine and $AlCl_3$, the major product is C_6H_5Br and not C_6H_5Cl. Draw a mechanism for this reaction to explain this observation.

(c) Explain whether you would expect either BBr_3 or NBr_3 to act as a catalyst for the bromination of benzene.

Use the Periodic Table.

....

Answer

(a) In $AlBr_3$: Al atom $3s^2\ 3p^1$ = 3 electrons in the outer shell
        One more from each Br   = $\underline{3}$ electrons
                Total   = $\underline{6}$ electrons

The Al is a pair short of the 8 needed for a full outer shell, so is 'electron-deficient'.

(b) Mechanism:

$$AlCl_3 \quad + \quad Br_2 \quad \rightleftharpoons \quad [\,AlCl_3Br\,]^- \ Br^+$$

We would need Cl_2 to make Cl^+ to make C_6H_5Cl, and there is no Cl_2.

(c) BBr_3: B atom $2s^2\ 2p^1$ = 3 electrons in the outer shell
    One from each Br   = $\underline{3}$ electrons
              Total   = $\underline{6}$, still electron-deficient,
so it could act as a catalyst like $AlBr_3$.

NBr_3: N atom $2s^2\ 2p^3$ = 5 in the outer shell
    One from each Br   = $\underline{3}$
             Total   = $\underline{8}$, a full shell,
so NBr_3 is not expected to act as a catalyst in the way.

Even if you made ClBr by

$$[\,AlCl_3Br\,]^- \ + \ Br^+$$
$$\updownarrow$$
$$AlCl_2Br \ + \ ClBr$$

the ClBr is likely to give Br^+ by the back reaction rather than Cl^+ (look at the electronegativities of Cl and Br).

H & P section 5.8 (p 71)

___

Question 16

(a) Draw a mechanism for the nitration of benzene using mixed concentrated sulphuric and nitric acids.

(b) If a large amount of $KHSO_4$ were added to the mixture, would you expect the nitration rate to be faster, slower or unchanged? Explain your answer.

(c) Assuming that the reaction of the electrophile with benzene is the rate-limiting step, draw an energy profile for the reaction. Label the activation energy E_a and draw structures for any intermediates.

Answer

These equilibria generate the electrophile NO_2^+.

(a)

$$H_2SO_4 + HNO_3 \rightleftharpoons HSO_4^- + H_2\overset{+}{N}O_3$$

$$H_2\overset{+}{N}O_3 \rightleftharpoons H_2O + \overset{+}{N}O_2$$

$$H_2O + H_2SO_4 \rightleftharpoons H_3\overset{+}{O} + HSO_4^-$$

overall $2\,H_2SO_4 + HNO_3 \rightleftharpoons 2HSO_4^- + H_3O^+ + NO_2^+$

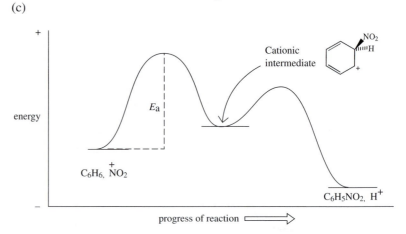

See La Chatelier's principle

(b)   A large increase in $KHSO_4$ concentration might be expected to push the first equilibrium to the left, reducing the concentration of HNO_3^+, which reduces the concentration of NO_2^+ and so slows down the reaction.

The cationic intermediate is stabilised by resonance delocalisation

(c)

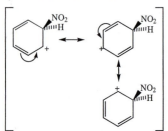

which can also be written as

H & P section 5.8 (p 70)

Question 17

(a)   When $AlCl_3$ is mixed with isotopically labelled chlorine, $^*Cl_2$, scrambling of the * label occurs between the two compounds. Explain this observation.

(b)   Similar isotopic scrambling occurs between CH_3COCl^* and $AlCl_3$. Suggest an explanation for this.

Look up the mechanism for the halogenation of benzene.

(c)   Suggest a mechanism for the reaction between C_6H_6, CH_3COCl and $AlCl_3$.

....

Answer

(a) A series of equilibria is set up.

$$AlCl_3 + \overset{*}{Cl}-\overset{*}{Cl} \rightleftharpoons \left[Cl-\overset{\underset{\displaystyle Cl}{|}}{\underset{\underset{\displaystyle Cl}{|}}{Al}}-\overset{*}{Cl} \right]^- \overset{*}{\overset{+}{Cl}} \rightleftharpoons AlCl_2\overset{*}{Cl} + Cl-\overset{*}{Cl} \rightleftharpoons \left[Cl-\overset{\underset{\displaystyle {}^*Cl}{|}}{\underset{\underset{\displaystyle Cl}{|}}{Al}}-\overset{*}{Cl} \right]^- \overset{+}{Cl} \quad etc.$$

(b)

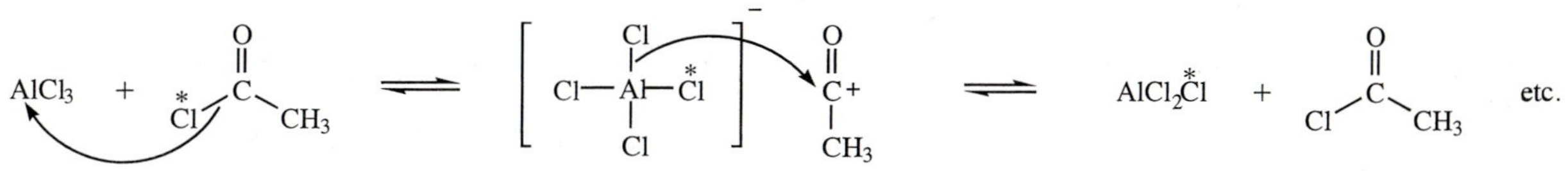

(c)

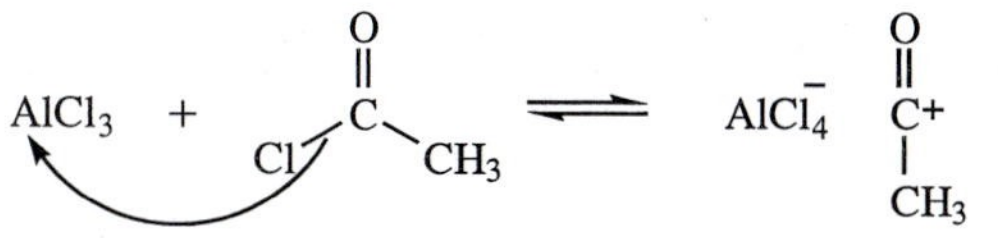

The charge in this acyl carbocation is stabilised by resonance delocalisation.

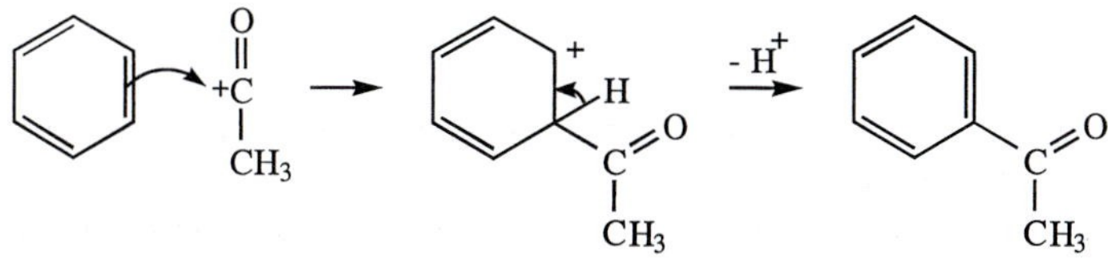

$$\left[CH_3\overset{+}{C}=\ddot{O}: \longleftrightarrow CH_3C\equiv\overset{+}{O}: \right]$$

This is an example of a Friedel–Crafts acylation reaction.

H & P section 5.8 (p 69)

6 Reactions with radical intermediates

Questions by topic

A: 1, 2, 3, 4, 5,

B: 6, 7, 8

6.1 What you need to read about or revise for this chapter

A Formation and nature of radicals; chain reactions; use of fish-hook arrows. Halogenation of hydrocarbons: mechanism.

B Radical polymerisation of alkenes: mechanism.

6.2 Worked examples

Question 1

(a) Write the overall equation for the chlorination of ethane to give chloroethane under ultraviolet (UV) irradiation. Suggest a mechanism for this reaction, and explain why a very small quantity of butane is also made.

(b) Draw the mechanism as a radical flow diagram.

For radical flow diagrams see H & P p 77.

Answer

(a) Overall

$$CH_3CH_3 + Cl_2 \longrightarrow CH_3CH_2Cl + HCl$$

Radical chain mechanisms involve initiation, propagation and termination steps.

Single bonds between two electronegative elements are relatively weak: see also O–O and N–N. One reason for this is the interelectron repulsion between the nonbonded pairs of electrons on adjacent atoms.

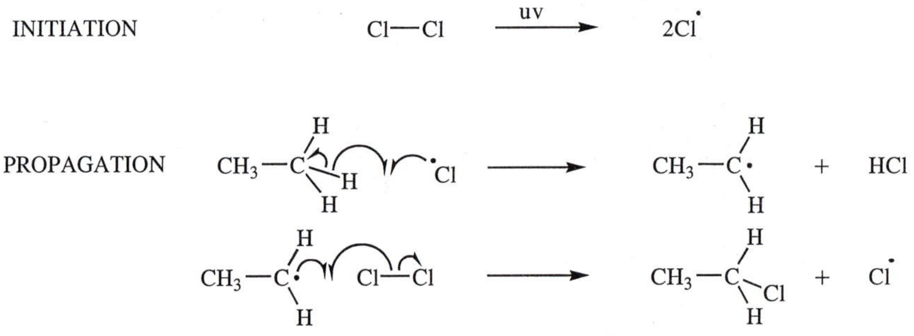

The Cl• goes back into the first stage of the propagation again to continue the chain.

TERMINATION   combination of any two radicals:

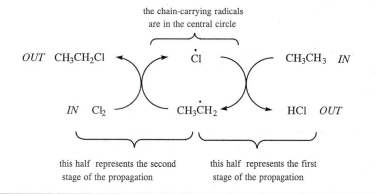

The propagation chains are long. There are few termination reactions, and only some of these give butane—so the yield of butane is very low.

(b)   Radical flow diagram.

the chain-carrying radicals
are in the central circle

OUT CH_3CH_2Cl ← → $\dot{C}l$ → ← CH_3CH_3 IN

IN Cl_2 → $CH_3\dot{C}H_2$ → ← HCl   OUT

this half represents the second
stage of the propagation

this half represents the first
stage of the propagation

H & P section 6.2 (p 75)

Question 2

(a)   Write the propagation steps for the chlorination of
(i) CH_2Cl_2 to $CHCl_3$; (ii) $CHCl_3$ to CCl_4.
(b)   Draw radical flow diagrams for these reactions.

....

Answer

(a)   (i) For CH_2Cl_2 to $CHCl_3$

$$HC-H \quad \cdot Cl \longrightarrow HC\cdot \;+\; H-Cl$$

Remember that if there is an odd number of electrons on one side of the equation, there must also be an odd number on the other side to balance.

$$H\dot{C} \quad Cl-Cl \longrightarrow HC-Cl \;+\; Cl\cdot$$

In the propagation steps, the last radical must be the same as the first so that a chain is formed.

(ii) For $CHCl_3$ to CCl_4

$$Cl_3C-H \quad \cdot Cl \longrightarrow Cl_3C\cdot \;+\; H-Cl$$

$$Cl_3C\cdot \quad Cl-Cl \longrightarrow Cl_3C-Cl \;+\; Cl\cdot$$

CHLOROFORM (CHCl$_3$) was one of the first general anaesthetics to be discovered, but it has serious side-effects. Modern anaesthetics are better and less toxic.

(b)   For a(i)

a product out   CHCl$_3$ ← $\dot{C}l$ → CHCl$_2$ *a starting material in*

a starting material in   Cl$_2$ → $\dot{C}HCl_2$ → HCl   *a product out*

overall equation: Cl$_2$ + CH$_2$Cl$_2$ = HCl + CHCl$_3$

(b)   For a(ii)

CCl$_4$ was used for dry-cleaning of clothes but has now been replaced by less toxic compounds.

CCl$_4$ ← $\dot{C}l$ → CHCl$_3$

Cl$_2$ → $\dot{C}Cl_3$ → HCl

CHCl$_3$ + Cl$_2$ = CCl$_4$ + HCl

H & P section 6.1, 6.2 (p 75)

Question 3

dibenzoyl peroxide

(a)   Explain why 2-chloropropane is the main product of electrophilic addition of HCl to propene in the dark.

(b)   If the reaction is carried out in the presence of peroxides, dibenzoyl peroxide for example, the product is largely 1-chloropropane.
  (i)   Writing the initiator as R$^\bullet$ suggest a mechanism for this radical addition of HCl to propene.
  (ii)   Draw a radical flow diagram for the propagation steps in (i).

The first step involves R$^\bullet$ reacting with HCl to form Cl$^\bullet$.

....

Answer

(a)   This is a polar reaction.

The secondary carbocation is more stable than the primary alternative, CH$_3$CH$_2$CH$_2^+$.

CH$_3$CH=CH$_2$ →   CH$_3\overset{+}{C}HCH_3$

H$\overset{\frown}{C}$l   Cl$^-$

CH$_3\overset{+}{C}HCH_3$ →   CH$_3$CHCH$_3$
                                                      |
Cl$^-$ Cl

(b)   This is a radical reaction. It goes *via* the more stable secondary radical.

The Cl attaches itself first, forming the secondary alkyl radical which is more stable than the alternative primary radical   CH$_3\dot{C}HCH_2$
                                                                                                                                                                  |
                                                                                                                                                                  Cl

(i)   R$^\bullet$ + HCl → RH + Cl$^\bullet$
       Cl$^\bullet$ + CH$_3$CH=CH$_2$ → CH$_3$C$^\bullet$HCH$_2$Cl
       CH$_3$C$^\bullet$HCH$_2$Cl + HCl → CH$_3$CH$_2$CH$_2$Cl + Cl$^\bullet$
       And so the chain continues.

(ii)   CH$_3$CH$_2$CH$_2$Cl ← $\dot{C}l$ → CH$_3$CH=CH$_2$

HCl →   CH$_3\dot{C}HCH_2Cl$ ←

H & P section 6.2, 6.5 (pp 60, 79)

Question 4

(a) Suggest structures for the isomeric products **A** and **B** below, giving mechanisms to support your ideas.

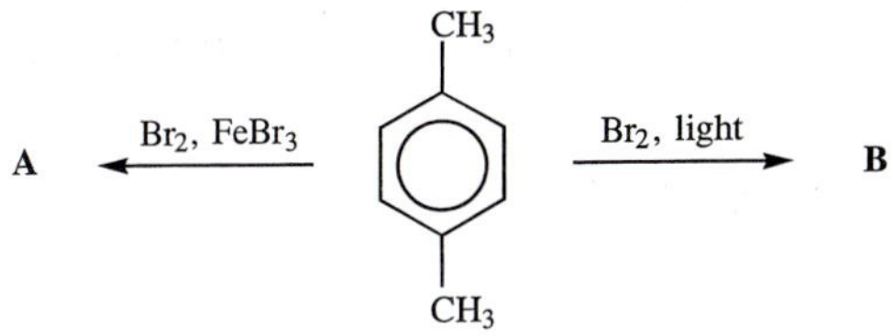

(b) Suggest and explain a simple chemical test which would enable you to differentiate between a sample of **A** and a sample of **B**.

For the test you should give reagents and conditions, then results (observations) for both compounds.

....

Answer

(a) **A**: (heterolytic) electrophilic aromatic substitution.

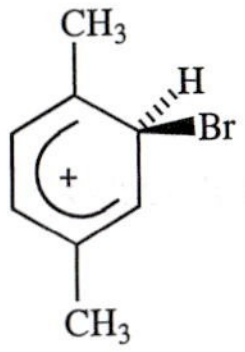

$$Br_2 + FeBr_3 \rightleftharpoons \overset{+}{Br} + \overset{-}{FeBr_4}$$

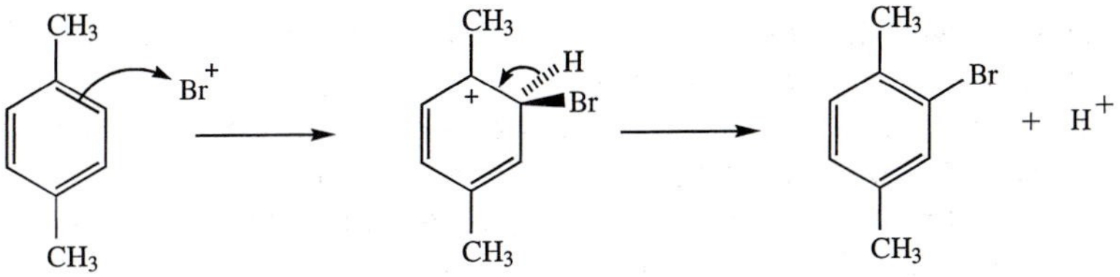

A, C_8H_9Br

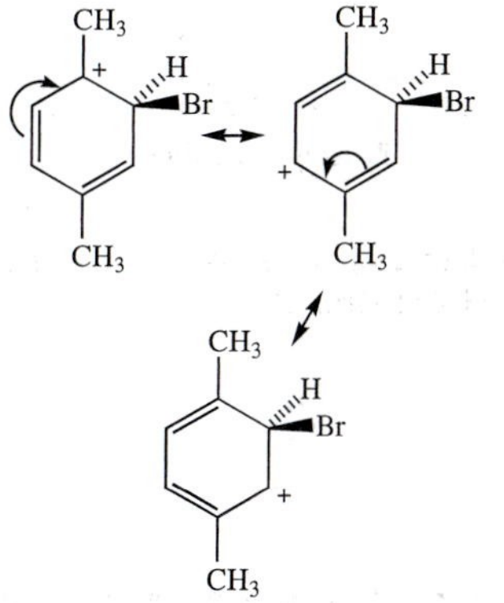

$$\overset{+}{H} + \overset{-}{FeBr_4} \rightleftharpoons HBr + FeBr_3$$

B: (homolytic) radical chlorination of the side-chain.

INITIATION $\qquad Br_2 \xrightarrow{uv} 2Br^\bullet$

PROPAGATION

B, C_8H_9Br

The intermediate cation is delocalised.

or in more detail:

This radical is delocalised and stabilised.

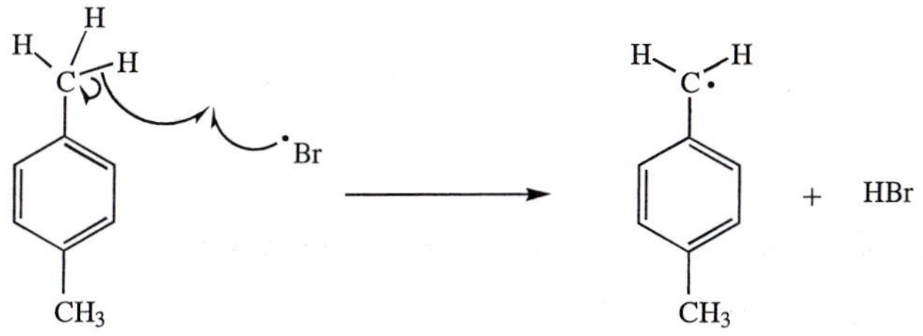

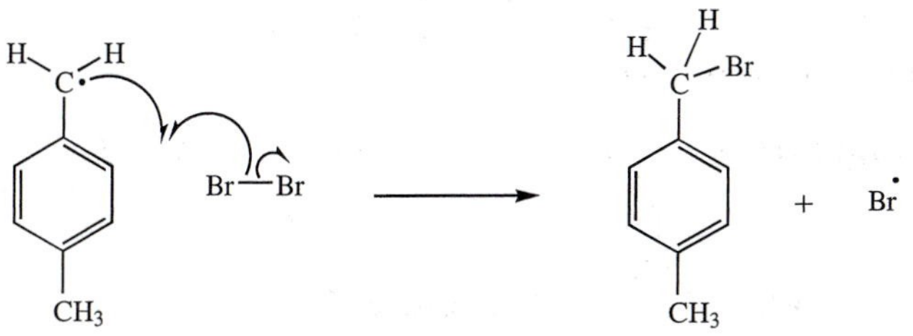

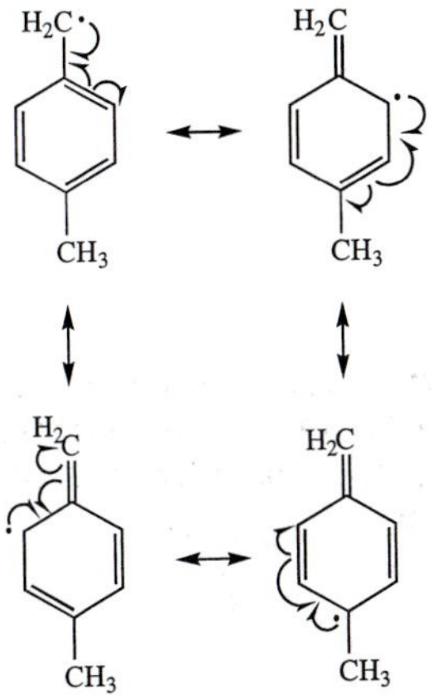

Radicals formed by loss of an H atom from the aryl ring cannot be delocalised.

TERMINATION   for example:

(b)

A is an aryl bromide and is very *unreactive* to nucleophiles.

B behaves like a bromo-alkane and is very reactive towards nucleophiles.

A mixed solvent is suggested because A and B are unlikely to be very soluble in water.

It is essential to acidify after cooling because AgNO$_3$ would form a precipitate of Ag$_2$O with any excess NaOH. HCl and H$_2$SO$_4$ form insoluble silver salts so these should not be used to acidify the solution.

Test   Warm the sample of **A** or **B** with aqueous alcoholic NaOH solution. Cool, acidify with HNO$_3$ and add AgNO$_3$(aq).

Observations:

Sample **A** will show no change; it may not even dissolve.

Sample **B** will give a yellowish precipitate of AgBr.

H & P sections 4.10, 5.8, 6.2 (pp 56, 69, 77)

$$NaBr_{(aq)} + AgNO_3{}_{(aq)} \longrightarrow AgBr_{(s)} + NaNO_3{}_{(aq)}$$

Question 5

ISOFLUORANE

Chlorofluorocarbons (CFCs), such as CCl$_2$F$_2$, have been used as refrigerants and aerosol propellants, but are now banned in the UK. More complex compounds, such as isofluorane, are still used as anaesthetics.

CFCs persist in the atmosphere until they reach the stratosphere, where the molecules are photolysed and reactions such as the following may occur.

The ClO$_2^{\bullet}$ molecule has the Cl between the two O atoms.

(i) $CCl_2F_2 \xrightarrow[\text{UV light}]{} {}^{\bullet}CClF_2 + Cl^{\bullet}$

(ii) $O_3 \xrightarrow[\text{UV light}]{} O + O_2$

(iii) $Cl^{\bullet} + O_3 \longrightarrow ClO^{\bullet} + O_2$

(iv) $ClO^{\bullet} + O \longrightarrow Cl^{\bullet} + O_2$

(v) $O_3 + ClO^{\bullet} \longrightarrow ClO_2^{\bullet} + O_2$

(a) Redraw reactions (i), (ii) and (v) using dot diagrams showing outer shell electrons only. State the type of bond breaking involved in (i).

(b) Which pair of reactions constitutes a chain reaction? Explain your answer and rewrite the reactions as a radical flow diagram.

(c) Suggest a reason for the continued allowed use of CFC anaesthetics, despite the ban on CFCs as refrigerants and aerosol propellants.

(d) Suggest reasons for the lack of reactivity of CFCs, to water for example, in the lower atmosphere.

Make sure that there is the same number of electrons on both sides of the equation.

CCl₄ is similarly unreactive.

....

Answer

(a) Reaction (i)

$$:Cl:C:F: \longrightarrow :Cl\cdot \; + \; \cdot C:F:$$

It is the weaker C–Cl bond which breaks, not C–F.

Homolytic bond breaking.

Reaction (ii)

18e   6e   12e

18 electrons on each side of the equation.

18e   13e   12e   19e

31 electrons on each side of the equation.

(b) $Cl^\bullet + O_3 \rightarrow ClO^\bullet + O_2$
$ClO^\bullet + O \rightarrow Cl^\bullet + O_2$

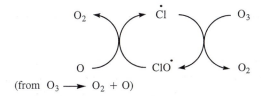

(from $O_3 \longrightarrow O_2 + O$)

Ozone is consumed in this cycle and the radicals are regenerated, which leads to some depletion of the ozone layer. The major route for ozone depletion occurs by a similar chain involving the dimer Cl_2O_4.

(c) Because the anaesthetics are very expensive, excess is recycled. Very little escapes into the atmosphere. Refrigerants and aerosols are used on a far greater scale and their release is much more difficult to prevent. In addition, isofluorane has a carbon atom bearing an H and a Cl, which means that it can be hydrolysed and degraded.

(d) The most likely reactions are nucleophilic substitution reactions in raindrops, either by a bimolecular (S_N2) or a unimolecular (S_N1) mechanism.

It is interesting that in contrast to CCl₄, SiCl₄ is easily hydrolysed. This may be because the greater size of Si means that this atom is more exposed to nucleophiles. More likely is that Si uses its 3d orbitals to receive initial attack by the nucleophile (H₂O) so that reaction proceeds by an addition–elimination mechanism (which is not possible for carbon).

H & P sections 6.1, 6.2 (p 75) also 4.2, 4.5 (pp 41, 45)

The small carbon atom is closely surrounded by large halogen atoms, so that access to the central carbon by a nucleophile (H_2O) is very restricted; and as the transition state would be even more crowded, the activation energy would be high. So S_N2 is difficult.

S_N1 reactions start with the formation of a carbocation by loss of a halide ion. This would put positive charge on a carbon already denuded of electrons by the strongly electron-withdrawing nature of the halogen atoms: this is unfavourable, so S_N1 is difficult too.

Question 6

Propene can be polymerised using dibenzoyl peroxide **A** as the initiator.

This involves homolytic bond breakage of the weak O–O bond (see Question 1, margin).

(a)   Writing the initiator as R•, draw a mechanism for the polymerisation of propene, and explain your choice of intermediates.

(b)   When heated or irradiated, **A** can break down to give CO_2 and $C_6H_5^•$ radicals. Suggest a mechanism for this process.

....

Answer

(a)

Poly(propene), also called POLY-PROPYLENE, is used for carpeting, upholstery and moulded objects like margarine tubs.

Termination: any combination of any of the radicals above. The first intermediate radical $RCH_2{}^\bullet CHCH_3$ is secondary, more stable than the alternative primary radical ${}^\bullet CH_2CHRCH_3$.

The next two intermediates are also secondary radicals.

(b)

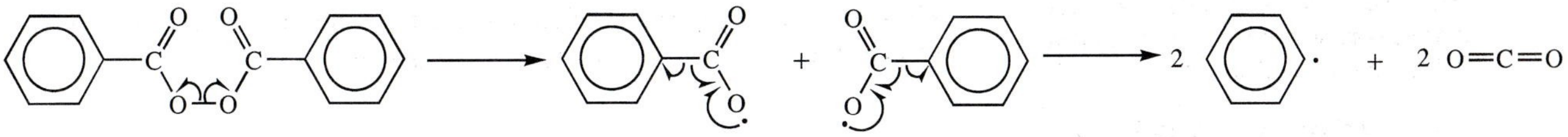

The weak O–O bond breaks homolytically.

The unpaired electron in $C_6H_5^\bullet$ is in an sp^2 orbital in the plane of the ring but at right angles to the ring's π system, so it cannot be delocalised with the π system.

H & P section 6.5 (p 79)

Question 7

The following polymers can be made by radical addition polymerisation. For each, suggest a structure for a suitable monomer.

(a)

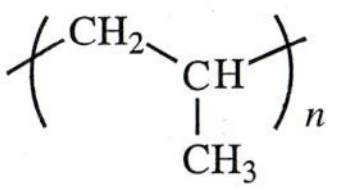

(b)

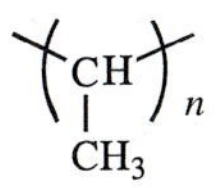

(c)

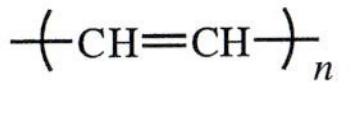

(d)

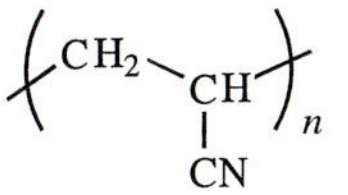

A monomer for *addition* polymerisation must have a multiple bond.

POLYACETYLENE (c) is a polymer which conducts electricity: all its double bonds are conjugated.

POLYACRYLONITRILE (d) is used to make knitwear, simulated fur fabric and carpets. Its copolymers with styrene and 1,3-butadiene are high-impact plastics used in fenders and crash helmets.

....

Answer

(a)

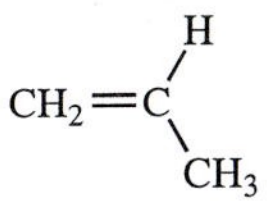

(b) 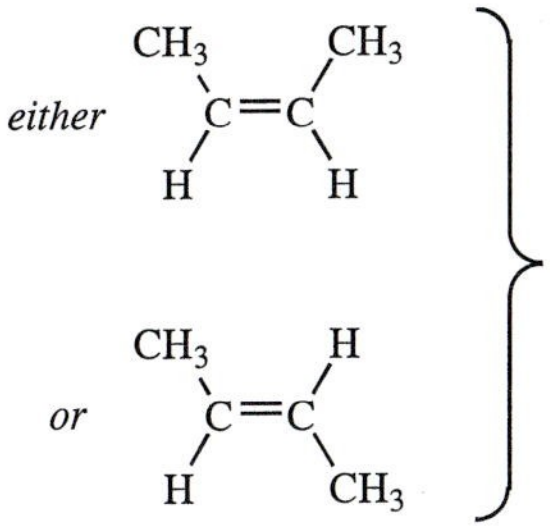 *either*

(c) $HC\equiv CH$

or

(d)

A quick way to do these is to 'fold' inwards the two outermost single bonds of the repeating unit to make an additional bond between the two-atom chain inside the brackets:

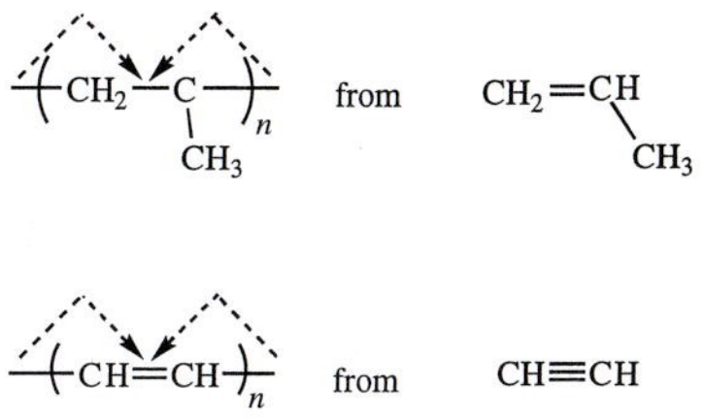

This **only** works for a **two-atom** chain inside the brackets. Always check your answer by working it forwards, from monomer to polymer.

H & P section 6.5 (p 79)

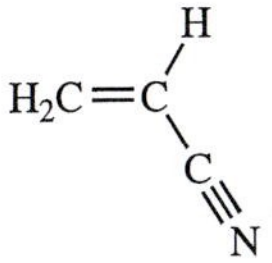

Radical polymerisation proceeds by addition of a radical to a multiple bond of the monomer. In (c) a double bond is retained in the polymer, so the monomer must have had a triple bond: an alkyne.

Question 8

$C_6H_5CH{=}CH_2$ and $CH_2{=}CHCOOCH_3$ can each form addition polymers.

(a) Draw the full structure of the repeating unit for both polymers.

(b) Which polymer would you choose for making bottles to carry concentrated KOH solution? Explain your answer, illustrating it with an equation.

Write down the functional groups in each polymer and consider possible reactions with KOH.

A quick way to do these for a monomer with one multiple bond: draw the multiple bond horizontally, with its substituents, and 'unfold' a bond outwards:

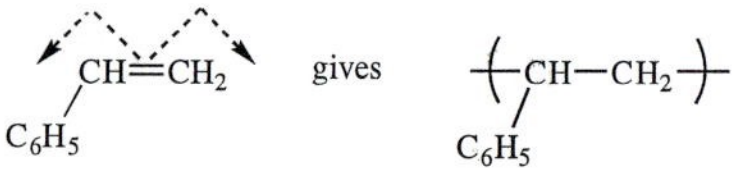

monomer repeating unit

POLYSTYRENE is used in packaging and in making toys and drinking cups. POLYACRYLATES are used in weatherproof coatings.

Answer

(a)

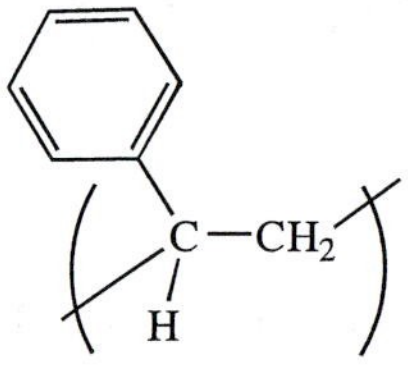

Polystyrene

and

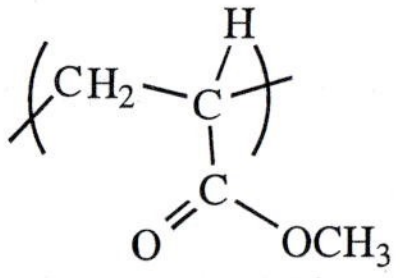

Poly(methyl acrylate)

NB. The repeating unit does *not* have the '*n*' outside the bracket.

(b) Polystyrene is better.

Reasons Poly(methyl acrylate), an ester, can be hydrolysed by the KOH solution to give a water-soluble polymer.

KOH will not react with the phenyl ring, nor with the hydrocarbon backbone of the polymer.

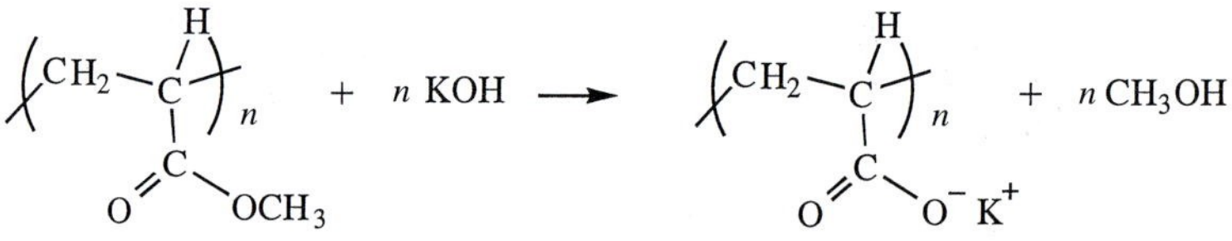

Gradually the KOH would 'eat' its way through the bottle. Polystyrene is inert to aqueous KOH, so should be suitable.

H & P section 6.5 (p 79)

7 Taking it further

7.1 What you need to read about and revise for this chapter

These questions are mixed, drawing on topics in all of the previous chapters. Several cover more than one topic, and you have no warning of which topics are involved. You need to know and revise all the previous chapters!

7.2 Worked examples

Question 1

The indicator phenolphthalein has different structures in acid (colourless) and in alkaline (deep red) solution in at pH > 10. Which of A or B is the red structure? Justify your choice.

A   B

Draw out the structures for **A** and **B** using ⬡ for the aromatic rings.

Look up *conjugation*.

A compound is coloured because it absorbs some wavelengths of visible light; the energy of the light absorbed corresponds to an energy gap between two electronic energy states of the molecule.

....

Answer

Coloured compounds have longer conjugated systems than colourless compounds. In **A**, all the six-membered rings are conjugated through the central sp^2 carbon. Check by looking for the long conjugated single–double–single–double . . . string of bonds in **A**.

  Note: In alkaline solution, acids lose protons to form anions; so **A** looks more likely anyway!

H & P ch 1 & 3 (pp 5, 33)

Question 2

When $NaNO_2$ reacts with iodoethane, a mixture of C_2H_5ONO and $C_2H_5NO_2$ can be formed. Use mechanisms to explain how these products arise.

A dot diagram of NO_2^- may be useful. NO_2^- is stabilised by resonance and the two O atoms are equivalent with equal N–O bond lengths:

....

Answer

NO_2^-:

So NO_2^- can act as a nucleophile through either its N lone pair or its O^- atom.

(The $-CO_2^-$ group is similarly delocalised.)

H & P Ch 1, 2 & 4 (pp 5, 20, 41)

or

Question 3

Ethanol can be oxidised to ethanal by heating with $K_2Cr_2O_7$ and aqueous sulphuric acid. A good yield of ethanal can be obtained by distilling the ethanal from the reaction mixture as it is formed. Two by-products of the reaction are ethanoic acid and ethyl ethanoate. The boiling points of these compounds are:

Ethanal: 21°C   Ethyl ethanoate: 77°C
Ethanol: 78°C   Ethanoic acid: 117°C.

Rationalise the pattern of these boiling points, illustrating your answer with diagrams.

....

Answer

Boiling points go up as intermolecular forces in the liquid get stronger, making it more difficult for molecules to escape from each other into the gas phase. The types of intermolecular forces—van der Waals, dipole–dipole and hydrogen bonding—can be deduced from the structure of each molecule.

Only O–H, N–H and F–H make strong hydrogen bonds; C–H does not.

Ethanal (M_r 44) has the lowest boiling point, 21°C. As well as weak van der Waals forces, the permanent dipole of the carbonyl group leads to slightly stronger dipole—dipole attractions between molecules.

Ethyl ethanoate (M_r 88) has a larger number of electrons than ethanal and so has stronger intermolecular van der Waals forces as well as the dipole–dipole attractions due to the carbonyl group. This raises its boiling point above that of ethanal.

Ethanol (M_r 46) can form hydrogen bonds with its hydroxyl group. These are stronger than dipole–dipole attractions so its boiling point is higher than that of ethanal.

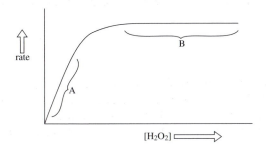

Ethanoic acid forms even stronger hydrogen bonds than ethanol, for example in chains and dimers. It has the highest boiling point, 117°C.

The O–H bond of ethanol is highly polarised

$$C_2H_5O\underset{\delta-}{}\!\!-\!\!\underset{\delta+}{H}$$

which contributes to its ability to form hydrogen bonds. The acidic O–H of ethanoic acid is even more highly polarised.

This is the ethanoic acid dimer.

H & P Ch 1 (p 15)

Question 4

The diagram shows how the rate of an enzyme-catalysed decomposition of hydrogen peroxide varies with changing peroxide concentration. All other conditions remain constant.

CATALASE is an iron-containing enzyme which catalyses the breakdown of H_2O_2, a potentially destructive agent in cells. The reaction involves the oxidation of one H_2O_2 to O_2 and the reduction of a second H_2O_2 to H_2O. Catalase is present in the liver.

(a) Explain why the reaction is described as first order with respect to H_2O_2 in region A but zero order in region B.
(b) Suggest a reason for the change in kinetics at high concentrations of H_2O_2, that is, from region A to region B.
(c) Which is the weakest bond in H_2O_2?
(d) Suggest a reason why organisms need a mechanism to destroy peroxides.

....

Answer

(a) A. The rate varies *linearly* with H_2O_2 concentration.
Rate $\propto [H_2O_2]$ or rate $= k\,[H_2O_2]$.
B. The rate is *independent* of H_2O_2 concentration, which corresponds to zero-order kinetics.
(b) In region B, the *enzyme* concentration has become the rate-limiting factor; the enzyme present reacts with H_2O_2 molecules as fast as it can.
(c) The O–O bond is weaker than the O–H bond. Interelectron repulsion between the nonbonded pairs on the O atoms is a contributing factor.

The active sites on the enzyme are saturated (all occupied). Another substrate molecule can only bind if an active site is vacated.

H & P Ch 2 & 6 (pp 18, 26, 81)

(d)   The weak O–O bond can break homolytically to produce two reactive, destructive radicals which could damage the cells.

$$R-O \cdots O-R \longrightarrow 2\ R\dot{O}$$

Question 5

The diagram shows some interactions of the side-chains in two sections of a protein which might be involved in maintaining its tertiary structure at about pH7.

The tertiary structure of a protein is the three-dimensional shape adopted by its polypeptide chains.

(a)   Identify the major type of attraction involved in each case. Draw a diagram to illustrate your answer to B.

(b)   Predict the effect on A, B and C of adding acid to bring the pH down from 7 to about 2.

Answer

Remember that for strong hydrogen bonding,

$$X \cdots H - Y$$

both atoms X *and* Y must be electronegative, e.g. F, O or N.

(a)   A: Ionic (coulombic) attraction between positive and negative ions.

B: Hydrogen bonding of a hydrogen atom between two electronegative atoms (oxygen).

C: van der Waals (hydrophobic) forces, due to temporary tiny dipoles caused by momentarily uneven electron distribution.

D: Covalent bonding.

Disulphide bridges are formed when bread dough is kneaded. They are important in achieving the correct viscoelastic properties in the dough.

(b)   A: The –COO$^-$ will be protonated to give –COOH, and the ionic attraction will be lost. Hydrogen bonding is still possible both directly and indirectly *via* water.

B: No change. Primary alcohols are not basic enough to be much protonated at pH2.

C: No change.

The pKa values for protonated alcohols are $\sim$ –2, so at pH2 the ratio RCH$_2$OH: RCH$_2$O$^+$H$_2$ is 1:10^4.

H & P Ch 1 & 3 (pp 15, 33, 36)

Question 6

CDCl$_3$ is commonly used as a solvent for proton nuclear magnetic resonance, because it has no protons to interfere with the signals you want to observe.

Account for the observation that when CHCl$_3$ is shaken with NaOH in D$_2$O, some CDCl$_3$ is obtained.

Answer

OH$^-$ is a base and the H of CHCl$_3$ is acidic; the $^-$CCl$_3$ anion is stabilised by the electron-withdrawing inductive effect of three Cls.

$$HO^- \quad H-CCl_3 \quad \rightleftharpoons \quad H_2O \quad + \quad {}^-CCl_3$$

The solvent is D_2O, so instead of picking up H from the few molecules of H_2O, $^-CCl_3$ will pick up D from the many molecules of D_2O to form $CDCl_3$.

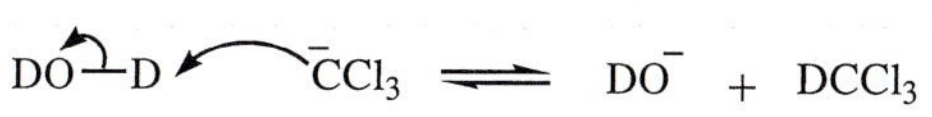

$$DO-D \quad {}^-CCl_3 \quad \rightleftharpoons \quad DO^- \quad + \quad DCCl_3$$

This H_2O will scramble with the D_2O; so will OH^-.

$$H_2O \quad + \quad D_2O \quad \rightleftharpoons \quad 2HDO$$

$$HO^- \quad + \quad D_2O \quad \rightleftharpoons \quad HDO \quad + \quad DO^-$$

H & P Ch 3 (p 37)

Question 7

Pick out the peptide link in the compounds zestril and captopril. Draw the structures of the amines and acids from which these peptide links could be made. How many different amino acids are involved?

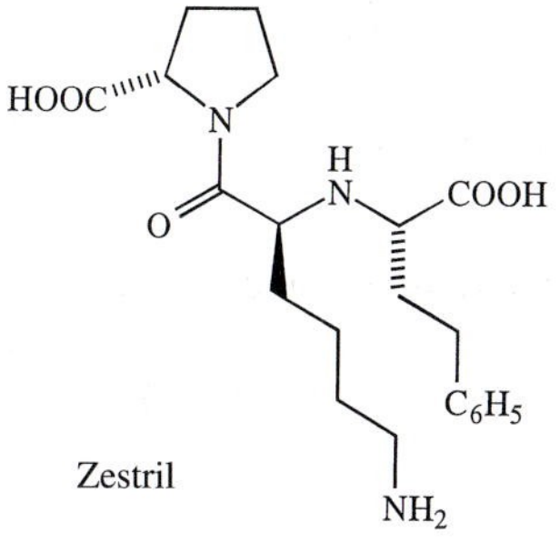

Zestril

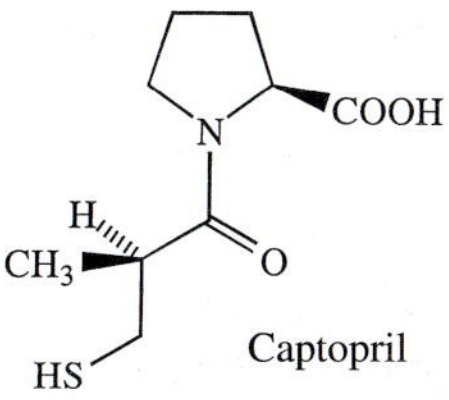

Captopril

ZESTRIL and CAPTOPRIL are used in the treatment of high blood pressure and heart failure. Both will bind to zinc ions and are inhibitors of angiotensin-converting enzyme.

Answer

Zestril

Amino acids

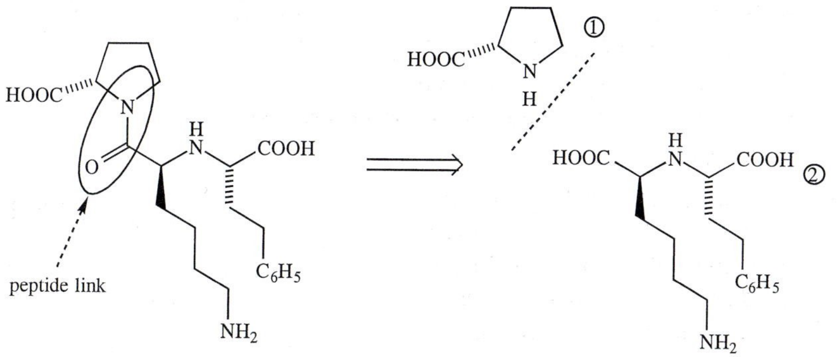

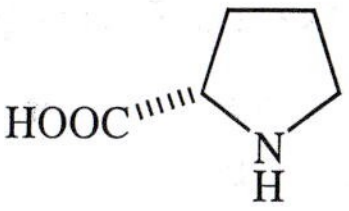

This is the essential amino acid proline.

Captopril

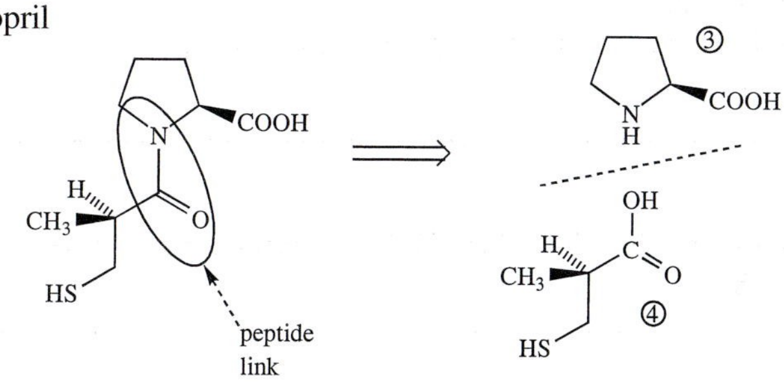

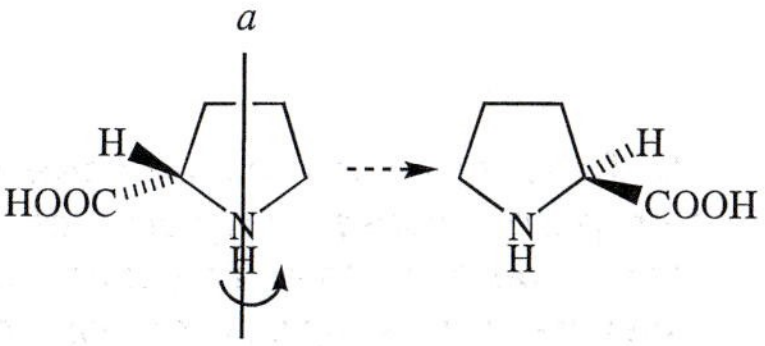

Only *two* amino acids are involved ① + ②.
① is the same as ③, and ④ is not an amino acid.

Rotate about axis *a*: ① ≡ ③

H & P Ch 1 & 4 (pp 8, 11, 14, 53)

Question 8

Suggest reagents and conditions for the conversion of $C_6H_5CH=CH_2$ into the following compounds. (More than one stage may be needed.)

Draw out the structures. Note the functional groups of the target compounds and try to relate these to the alkene starting material.

(a) $C_6H_5\overset{CH_3}{\underset{Br}{CH}}$

(b) $C_6H_5CHCH_2D$ with D

(c) C_6H_5CHO

(d) $C_6H_5\overset{CH_3}{\underset{OH}{CH}}$

(e) poly(phenylethene)

(f) $C_6H_5\underset{OH}{CHCH_2Br}$

(g) $C_6H_5\overset{H}{\underset{CN}{C}}{-}OH$

(h) $C_6H_5\underset{\overset{\|}{O}}{C}CH_3$

.....

Answer

The product is a bromoalkane; H and Br have been added to C=C.

(a)

by HBr addition to the C=C *via* the delocalised secondary C^+ ion, $C_6H_5\overset{+}{C}HCH_3$ (more stable than $C_6H_5CH_2\overset{+}{C}H_2$).

Two D atoms have been added. The chemistry of D_2 is very similar to H_2, so treat it in the same way.

(b)

by catalytic addition of deuterium (D_2) in the presence of Pd on C—just like hydrogenation.

The loss of a carbon atom is a pointer towards ozonolysis, which splits a C=C bond.

(c)

by ozonolysis of the double bond: O_3 followed by Zn + H_2O.

(d)

by hydration of the double bond using concentrated H_2SO_4 followed by water *via* the same cation as in (a).

$C_6H_5-C\overset{O}{\underset{O-O}{}}$ ... $C-C_6H_5$

Dibenzoyl peroxide

(e)

by acid-catalysed polymerisation in the presence of H^+ or a Ziegler–Natta catalyst (a Lewis acid) OR by radical polymerisation induced by an initiator such as dibenzoyl peroxide.

(f)

Initial reaction with C=C by the best electrophile, Br_2, is followed by nucleophilic attack on the positively charged intermediate by water (both the solvent and a nucleophile).

by addition of bromine water to the double bond, *via* the more stable intermediate.

(g)

see (c)

Best to work backwards here:

is a cyanohydrin, made from the carbonyl group

by nucleophilic addition of HCN using KCN/H_2SO_4.

(h)

see (d)

Redox interconversions, like this ketone to secondary alcohol and back, are often useful in synthesis.

by oxidation of the secondary alcohol to the ketone by $Na_2Cr_2O_7/H^+$ or by CrO_3. Gentle oxidation by $KMnO_4$ may also work.

H & P Ch 4, 5 & 6 (pp 48, 59, 61–66, 78, 79)

Question 9

Sulphuryl dichloride, SO_2Cl_2, can be used for the photochemical chlorination of methylbenzene. Suggest a mechanism for this process, and give the overall equation. Explain your choice of major organic product.

'Photochemical' often indicates a radical mechanism.

....

Answer

INITIATION

$$\xrightarrow{\text{u.v. light}} SO_2 + 2\overset{\bullet}{C}l$$

PROPAGATION

$$C_6H_5CH_3 + \overset{\bullet}{C}l \longrightarrow C_6H_5\overset{\bullet}{C}H_2 + HCl$$

$$\longrightarrow C_6H_5CH_2Cl + SO_2 + \overset{\bullet}{C}l$$

TERMINATION e.g.

$$C_6H_5\overset{\bullet}{C}H_2 + \overset{\bullet}{C}l \longrightarrow C_6H_5CH_2Cl$$

The reaction is terminated by the combination of any two radicals involved in the propagation stages.

Equation :

$$C_6H_5CH_3 + SO_2Cl_2 \longrightarrow C_6H_5CH_2Cl + SO_2 + HCl$$

The fastest reaction goes through the most stable intermediate radical to give the major product.

The possible radicals made by Cl• removing a hydrogen atom from $C_6H_5CH_3$ are:

| A | B | C | D |

removal of
an H from CH_3

removal of an aryl C-H

Only the first, **A**, is stabilised by delocalisation with the π system so this is the most stable radical intermediate.

A:

None of the other radical intermediates (B, C or D) is delocalised in this way.

So the product is $C_6H_5CH_2Cl$, derived from the most stable radical **A** by the lowest-energy process.

H & P Ch 6 (p 75, 77)

Question 10

Compound **A**, methyl *cis, cis*-9,12-octadecadienoate, occurs naturally in the linseed oil and is an important constituent of paint.

> Linseed oil comes from flax. It is also used in making linoleum floor-coverings.

A $CH_3(CH_2)_4CH\!\!=\!\!CHCH_2CH\!\!=\!\!CH(CH_2)_7COOCH_3$

(a) Explain the way **A** is named and draw its structure to show all multiple bonds and their stereochemistry.

(b) When paint dries in air, it first loses weight as the liquid added to make it spread evaporates. The next hardening process is accompanied by a gain in weight.
   (i) Suggest a possible chemical explanation for the second process.
   (ii) Devise and explain a simple experiment to support your suggestion.

> If there is a terminal functional group, like the carboxylate here, then counting starts at its carbon atom. So 2-methyl butanoic acid is
>
> $$CH_3CH_2\overset{2}{C}H\overset{1}{C}OOH$$
> $$| \atop CH_3$$
>
> $$\left(\text{and } not \quad CH_3CHCH_2COOH \atop | \atop CH_3 \right.$$
> $$\left. \text{which is 3-methyl butanoic acid} \right)$$

Answer

(a) 'Octadeca' means 18 carbons in a chain as the skeleton.
'Methyl . . . oate' means the methyl ester of the terminal carboxylic acid.
'9,12 . . . dien . . .' means that there are two C=C double bonds, starting at C9 and C12. So far we have this:

Cis, cis means that both C=Cs have the carbon substituents on the *same* side. Fill in everything else with H atoms to get

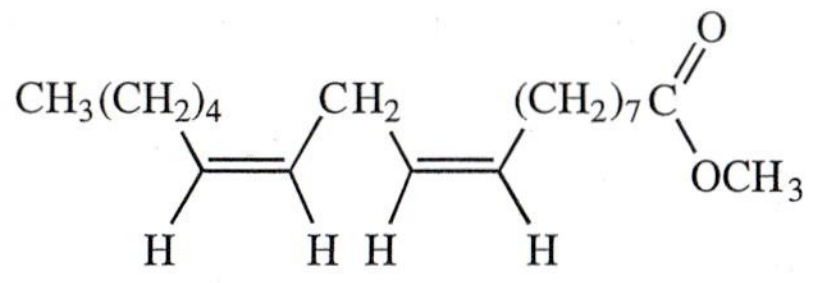

or as a skeletal diagram:

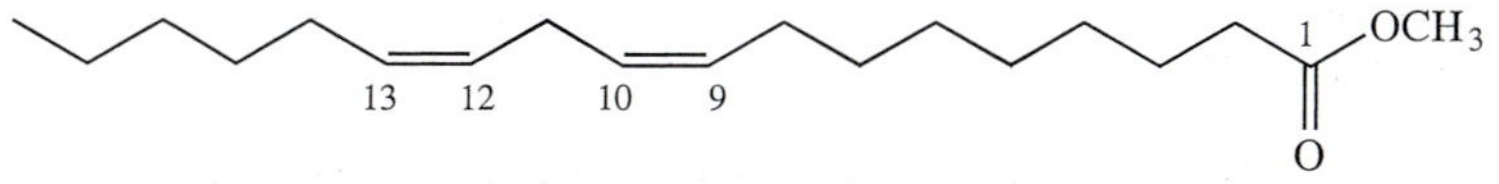

(b) (i) Cross-linking with oxygen could occur which would account for the weight gain. For example:

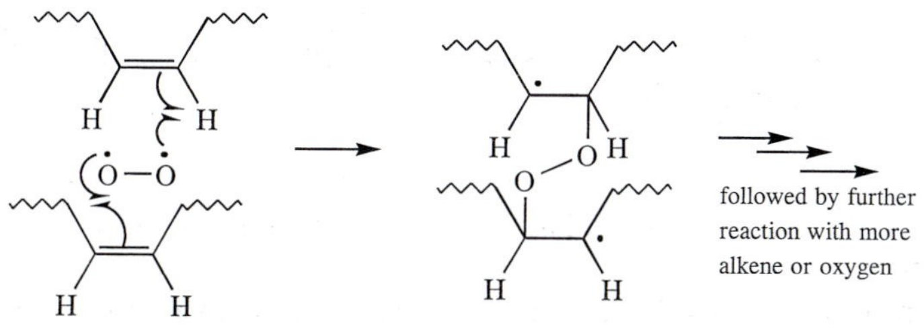

followed by further
reaction with more
alkene or oxygen

Cross-linking reduces the flexibility of the chains and so hardens the polymer. Natural rubber is cross-linked by sulphur (vulcanisation) to make it harder.

Oxygen is shown here as the diradical Ȯ–Ȯ. It can exist as either this or the electron-paired O=O.

(ii) Two possible experiments:
   1. See if the paint hardens and gains weight in the *absence* of O_2, that is, under a nitrogen atmosphere or in a vacuum. Alternatively, do the same experiment under pure O_2, not air; this time it should harden faster.
   2. Add a radical initiator, such as dibenzoyl peroxide. This should accelerate the hardening by polymerisation.
   The experiments in 1 seem to give better evidence.

Alternatively, the reaction could be stopped or slowed down by an inhibitor, such as a phenol, which combines with radicals to form very stable radicals which are unable to continue the chain. Several types of phenol are used as antioxidants in food.

H & P Ch 1 & 6 (pp 7, 13, 79)

Question 11

Wijs' reagent, a solution of iodine monochloride, ICl, is used to provide a quantitative measure of the extent of unsaturation in fats and oils.
(a) Write the equation for the reaction of iodine monochloride with ethene and suggest a mechanism.
(b) (i) Suggest a mechanism for the addition of iodine monochloride to propene, explaining in detail the relative positions of the two halogen atoms in the product.
   (ii) The product of the reaction of iodine monochloride and propene does not rotate the plane of plane polarised light. Explain why this is so.
(c) Iodine monochloride can also be used to halogenate some arenes. For example, 2-aminobenzoic acid can be converted into 2-amino 4-iodobenzoic acid in 70% yield by a solution of iodine monochloride. Suggest a mechanism for this reaction.

Look at the mechanisms for reactions with Br_2 as a guide for all of these. Think about any polarisation in iodine monochloride.

Answer

(a) $H_2C=CH_2 + ICl \rightarrow CH_2ICH_2Cl$

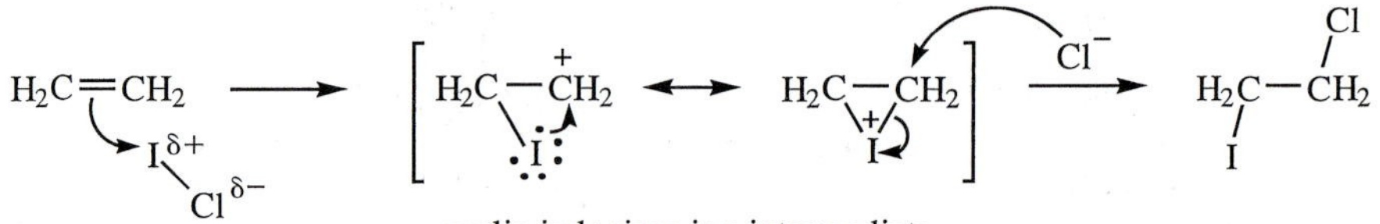

cyclic iodonium ion intermediate

electronegativity Cl > I
so I is $\delta+$, Cl is $\delta-$.

(b) (i)

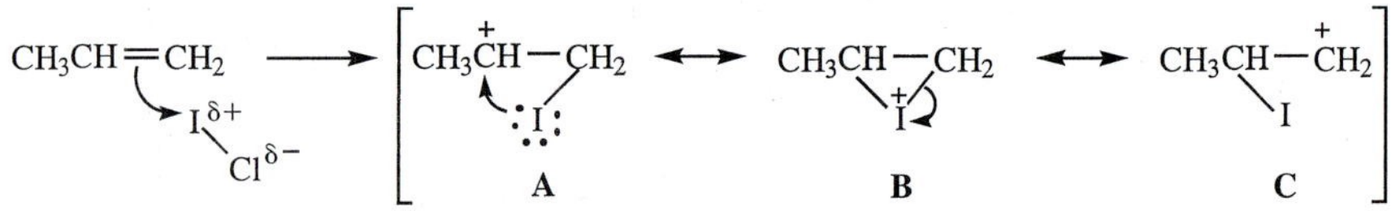

cyclic iodonium ion intermediate

Contributing structure **A**, a secondary carbocation, is more stable than **C**, a primary carbocation, so **A** is a better contributor to the overall structure than **C**. Hence the actual intermediate has more positive charge on the secondary carbon and the nucleophile adds there.

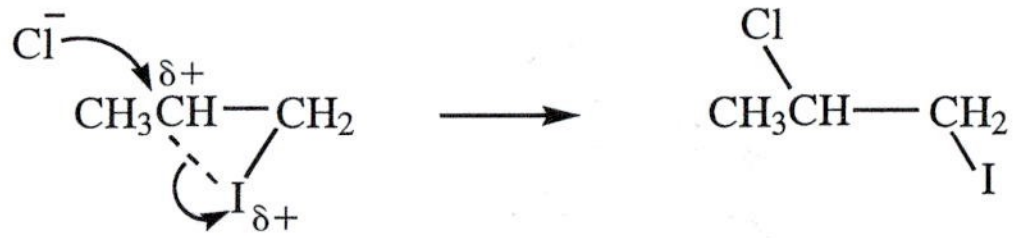

The major product is 1-iodo 2-chloropropane, $CH_3CHClCH_2I$, and not 1-chloro 2-iodopropane, CH_3CHICH_2Cl.

(ii) The product is chiral: C2 has four different groups, CH_3, Cl, H and CH_2I. None of the starting materials or reagents is optically active, so the product will not be optically active either: it will be a 50:50 mixture of the two optical isomers. These arise by two equally possible pathways.

Either

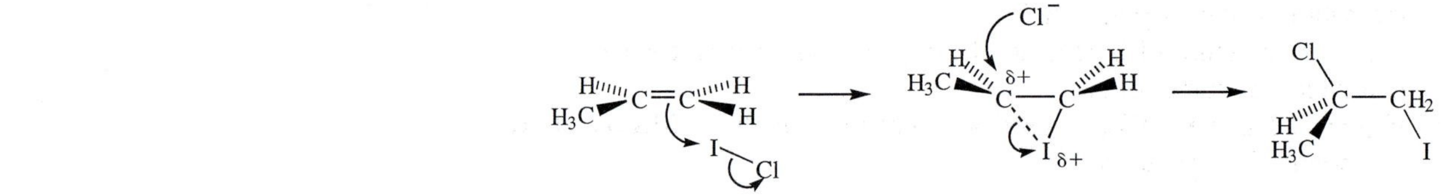

or ICl approaches the flat alkene from above instead of below:

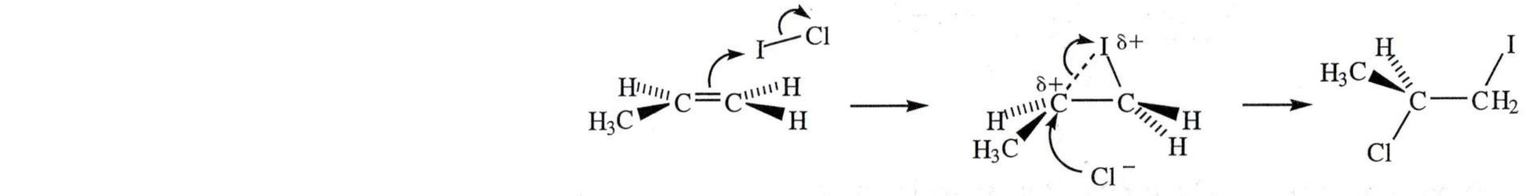

The two products drawn are non-superimposable mirror images of each other. The 50:50 mixture does not rotate the plane of plane polarised light.

(c) Reaction:

COOH, NH₂ + ICl ⟶ COOH, NH₂ (with I) + HCl

If there is no base present, the product will be the salt

COOH, $\overset{+}{N}H_3$ Cl^- (with I)

Mechanism:

COOH, NH₂ (with I–Cl) ⟶ COOH, NH₂, I, H, + (Cl⁻) ⟶ COOH, NH₂, I, H⁺

HCl

The delocalised cationic intermediate is further stabilised by the nonbonded pair on N.

[COOH, :NH₂, I, +, H ⟷ COOH, $\overset{+}{N}H_2$, I, H]

H & P Ch 1 & 5 (pp 14, 62, 71)

Question 12

Poly-L-leucine, PLL, is a synthetic polymer of the natural amino acid L-leucine. It can act as a catalyst in the oxidation of **A** to **B**.

H, CH₂CH(CH₃)(CH₃), C, ⁻OOC, $\overset{+}{N}H_3$

L - leucine

A ⟶ (H₂O₂, NaOH, PLL catalyst) ⟶ B

(a) Draw the dipeptide made from two leucine molecules, showing clearly the nature of the 'peptide link'.

(b) Draw the structure of D-leucine, the optical isomer of L-leucine.

(c) Draw the *cis* isomer of **A**.

(d) If poly-D-leucine, PDL, is used as a catalyst for the oxidation of **A**, compound **C** is produced.

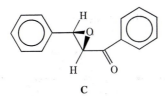

C

Explain how **B** and **C** are related as stereoisomers.

Look up the genetic code.

(e)   Suggest a reason why it is possible to make PLL but not PDL with a bacterial culture using a poly-UUC messenger RNA.

....

Answer

(a)

The 'peptide link' is a substituted amide.

(b)   The optical isomer is the non-superimposable mirror image of L-leucine, which is D-leucine.

D-leucine   L-leucine

(c)   This is the *cis* isomer, with the two Hs on the same side of the double bond.

(d)   **B** and **C** are non-superimposable mirror images of each other.

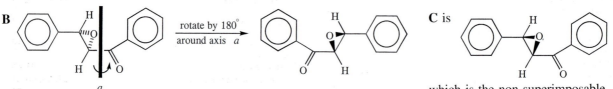

B   rotate by 180°
around axis *a*

C is

which is the non-superimposable mirror image isomer of **B**.

(e)   Most biological systems can only handle L-amino acids. UUC is the codon for L-leucine but will not work with D-leucine.

H & P Ch 1 & 4 (pp 13, 53)

Index

Functional Groups

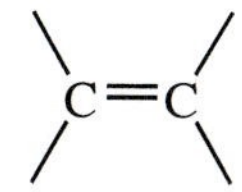

alkene

alkyne

arene or or

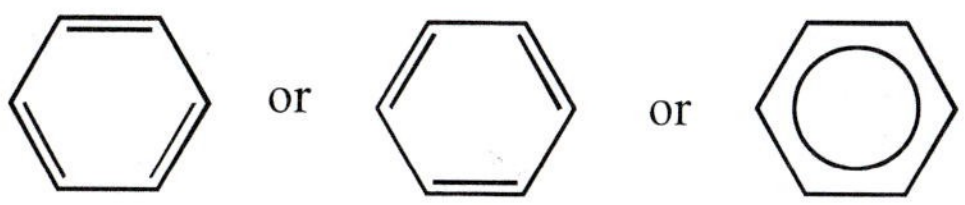

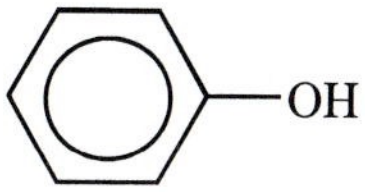

phenol

primary alcohol

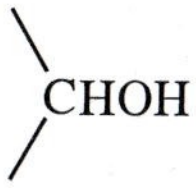

secondary alcohol

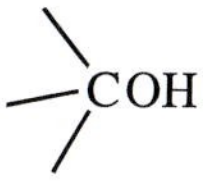

tertiary alcohol

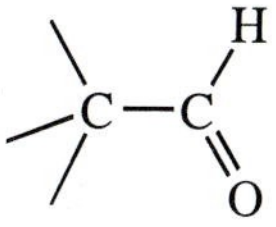

aldehyde

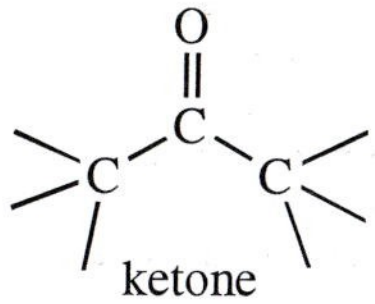

ketone

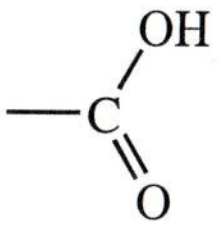

carboxylic acid

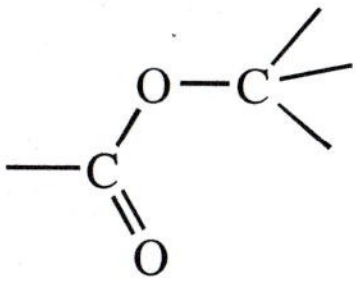

ester

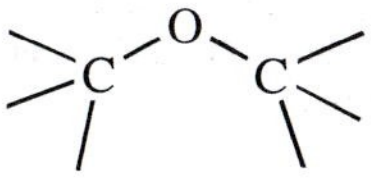

ether

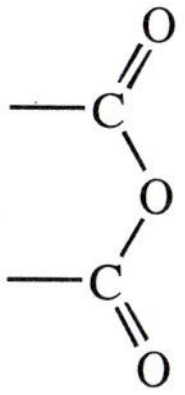

acyl anhydride
(or acid anhydride)

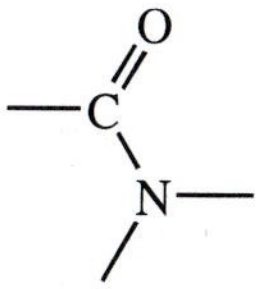

amide

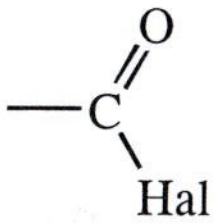

acyl halide
(or acid halide)
(Hal = F, Cl, Br or I)

nitrile